Sea Travels

Memoirs of a 20th Century Master Mariner

J. Holger Christensen
as told to Vaughn Sherman

EDMONDS, WASHINGTON

Copyright © 2012 by Vaughn Sherman

Published by Patos Island Press
705 Spruce Street
Edmonds, Washington 98020
www.patosislandpress.com

Library of Congress Control Number: 2012943903

ISBN: 978-0-9847225-1-8

Printed in the United States of America

Book and cover design
Jeanie James, www.ShorebirdMedia.com

All rights reserved. No part of this book may be reproduced or transmitted in any form or by any means, electronic or mechanical, including photocopying, recording or by any information storage and retrieval system without written permission from the author except for the inclusion of brief quotations in a review.

In memory of my father,

Neils Christensen

Table of Contents

Foreword

Chapter One: A Man from Fejø 1

Chapter Two: Early Days on Bainbridge Island 9

Chapter Three: *La Blanca* 17

Chapter Four: Working on Puget Sound 25

Chapter Five: Going Offshore 31

Chapter Six: *Griffson* 43

Chapter Seven: More Years on Deck 51

Chapter Eight: Loss of *La Blanca*,
    Building the *Hannah C.* 67

Chapter Nine: Surviving the Great Depression 79

Chapter Ten: The Strike and a Far East Experience 95

Chapter Eleven: An Historic Voyage 105

Chapter Twelve: Endings And Beginnings 117

Chapter Thirteen: Sailing into Wartime 131

Chapter 14: From Mate to Master 143

Chapter Fifteen: The *Stanley Griffiths* 153

Chapter Sixteen: The *Edmond Mallet* 171

Chapter Seventeen: Final Voyages 185

Afterword 191

About the Author 195

Master Mariner (n.)
DEFINITION FROM MERRIAM-WEBSTER ONLINE DICTIONARY

1: a captain of a merchant ship
2: an experienced and skilled seaman certified to be competent to command a merchant ship

The term "Master Mariner" is used variously in different countries, but in U.S. coastal waters, it means a license for commanding a vessel of any gross tonnage on any of the world's oceans. It is the highest-level marine license attainable.

FOREWORD

THIS STORY had its genesis in a 1984 trip to Europe planned with my late wife, Eunice Sherman. Eunice and I had lived nine years in Denmark and Sweden while I was serving with the Central Intelligence Agency. We wanted to visit friends there and in France, and to travel to London to take in some plays.

We were close to my uncle, J. Holger Christensen (he disliked his given name and used only the initial), who was born of Danish parents. When Holger learned of our plans he asked if it would be possible for him to go along. He had never visited Denmark and especially wanted to visit his mother's birthplace, the Baltic island of Bornholm, so we invited him to travel with us. Since we had never been to Bornholm, adding this to the itinerary made it a welcome side trip.

It was a wonderful time for all three of us. Holger was a big, tough, retired ship's captain, who in those days of retirement treated everyone with such charm that he was loved by all. That was also true during his sailing days unless he was dealing with a seaman or junior deck officer who disobeyed orders, or a troublesome sort in some bar in a distant port. Some of that dealing was with his fists.

In Sweden and Denmark we traveled by rented car, giving us many hours together without much to do. As it happened, those hours were filled with Holger's recounting of his life experiences. Like many men of the sea he was a natural storyteller. He told us about his upbringing on Bainbridge Island, located across from Seattle on Puget Sound, and of early years helping out on his dad's workboat. And about serving as a deckhand on "mosquito boats," small vessels that carried passengers between many Puget Sound ports. Later he told of progressing from seaman to captain on seagoing cargo ships calling at domestic and foreign ports.

I realized what a treasure trove of maritime information this was, covering the years from World War I to World War II and beyond. Holger and I made a deal. On our return from Europe we started a schedule that brought him once a week by car and ferry from his home on Bainbridge Island to mine in Edmonds, a town north of Seattle. He would bring along an audio tape recorded during the previous week, on which he had answered a list of questions I provided. First he would read and comment on what I had written during that time, then I would listen to his latest tape and ask questions while taking notes. Next I gave him a new list of questions to answer on tape during the following week. Then we'd go out to a fine lunch.

That occupied one of the more enjoyable years of my life. Sadly, my wife of many years, Eunice, died in 1985, about the time Holger and I finished the work. That slowed the job of pulling the information together. When I finished in 1988, Holger reviewed the completed manuscript and liked it. We decided to print the manuscript in a limited number of copies to be shared with family, friends, and interested parties such as museums and libraries. Unfortunately, Holger died in September 1988 before the printing was complete.

There have been many expressions of interest in this material through the years. Considering that, and the fact that modern technology now makes it possible to print attractive books at a reasonable cost, I've brought out this edition with additional material including maps of Holger's voyages.

I hope that those who saw the earlier work will find this book more interesting yet, and that new readers will enjoy traveling with Holger through the adventures of a career at sea.

Vaughn Sherman
Edmonds, WA
Fall 2012

CHAPTER ONE

A Man From Fejø

NEILS JULIUS CHRISTENSEN

The year is 1893, the season autumn. A young Danish sailor stands on the forward deck of a coastal schooner beating up Puget Sound towards a landing at Port Blakely on Bainbridge Island, across the sound from the little town of Seattle. As his vessel enters the port and the sailor works with the rest of the crew to prepare for anchoring, he looks at the tall-timbered island, senses the odors and sounds of a bustling port filled with big ships from around the world, takes a deep breath and realizes that he has finally "come home."

IT HAD TAKEN MY FATHER six years and countless sea miles served on ships sailing across many oceans to find home. Niels Julius Christensen was born May 16, 1874, at Østerby, a tidy little harbor town on the east side of the Danish Island of Fejø. His father was the postmaster on the island, a position with responsibilities quite different from those of modern day postmasters. In those days, the mail was delivered by boat to Vesterby, where my grandfather picked it up and delivered mail for Fejø by horse and wagon as he returned to Østerby. He then went on to the smaller islands in the vicinity of Fejø by boat, delivering mail for the residents there.

There were about 1200 people living on Fejø in the last half of the 19th century. Two career choices were open for young men born and raised there. One was to remain on the island, working the land. The other choice was to take to the sea as fishermen, as crewmen aboard the coastal freighters so vital to transportation then, or as deep-water sailors. There was a strong seafaring tradition in our family. Thorvald Christensen, my father's brother, owned his own small schooner and lost his life when his ship was wrecked in a storm near Fejø. My great-great-grandfather, Christian Pedersen Badike, was a seafaring man. His name appears on a sailing ship model that hangs in the church at Fejø, as one of the contributors to restoration of the model in 1854. And my postmaster grandfather had worked as a deep-sea diver before settling into the tranquil life of delivering mail.

Grandfather had wanted Dad to stay on the island as a farmer. But like so many before him, Dad ignored that advice and left Fejø at the young age of thirteen years to sail on coastal freighters as a cook. In later years, he recalled that he hadn't even boiled water before shipping out on those small schooners, calling at North Sea ports, but somehow he managed to learn to cook well enough not to be thrown overboard. At the same time he learned enough seamanship to be taken on full-rigged sailing ships as a seaman at the age of fifteen years. Following a sea-going career in those days was a dangerous job. I have an old letter from Denmark datelined "Fejø, March 14, 1894," addressed to "Dear Brother Niels". It reads in part:

"I am writing for Godfather, since he cannot see well enough to write by candlelight. No news to tell you, except that Thorvald is sailing from one port to the next and making money over by the Cape.

"I have to tell you that we have been worried about you during the winter, what with the bad weather, the likes of which we haven't seen in fifty years in Denmark, with so many maritime accidents, particularly out by the Skaw, where many lives have been lost and many ships destroyed. Please write us another letter before you leave that area." The writer was Anna Fredericksen, a cousin who had been taken in as an adopted daughter by my grandparents when her mother died.

After Dad's apprenticeship on coastal schooners, his first voyage as a seaman was out of Hamburg, on a course that took the ship down the coast of Africa, around the Cape of Good Hope and across the Indian Ocean to Sydney, Australia. He reminisced in later years about the beauty of that harbor and others, and the high standard of living and good wages for the working man. Melbourne was the last city they touched at, taking on a full cargo for the British Isles. The voyage home took 156 days. His most vivid memory of that trip was that "the food was pretty terrible!"

Dad's next trip was to take him around Cape Horn, the treacherous passage to the Pacific Ocean around the foot of South America. He sailed aboard a little Danish bark, bound for the west coast of Mexico with a cargo of wine. Then the ship sailed in ballast for the South Seas, where they loaded a full cargo of copra and mother-of-pearl. In the next couple of years, my father sailed twice more around Cape Horn. We don't know much about these trips, but the second must have been aboard a German bark, the *Maria Mercedes*. A discharge certificate, dated April 25, 1893 at Antwerp, Belgium, shows that he sailed aboard that ship as a seaman from May 9, 1892 to that date, that he "conducted himself in an "excellent" manner during that time, was also "eager to serve," and that he was discharged due to the "End of the Trip."

Not quite nineteen years old, with almost six years under sail, this young Dane from Fejø then went to Liverpool to ship out on a voyage that would take him around Cape Horn for the last time. The ship, a large barkentine, made one stop at Rio de Janeiro for water and then sailed non-stop to San Francisco. Dad left the barkentine at this point, but whatever he was looking for he didn't find in San Francisco. He found the port dirty, full of riff-raff and with bad living conditions. So he shipped out on the coastal schooner that took him for the first time to Bainbridge Island, the place he decided would be home.

Today a quiet little harbor with a few pleasure boats moored, the Port Blakely of the 1890's was one of the largest lumber ports in the world. After deciding that Bainbridge Island would be his home, Dad sailed the West Coast from there during the next three years on schooners carrying cargoes of lumber, returning with goods required by a fast-growing frontier area. In late 1896, Dad went into partnership with a Mr. Johnson to buy a 42-foot schooner. Together they prepared to take this little vessel on an extraordinary voyage. The destination was Attu, the last of the long chain of Aleutian Islands that stretches from Alaska almost to Japan. The purpose was to trap blue fox, an animal producing a pelt that was highly valued in those days. Besides the two partners, a third crew member was to be Mrs. Johnson, who was in a very pregnant condition by the time they departed Port Blakely.

In order to appreciate this adventure, the reader should know something about Alaska before the turn of the century. It's hard to imagine the size of the frontier, which had been bought less than thirty years earlier from Russia for only seven million dollars, or of the small population in the territory. According to the census of 1900 there were 29,500 Eskimos, Aleuts, and Indians living there, and about the same number of white Alaskans. The whites had numbered only 4,300 a decade earlier, but the discovery of gold had acted like a magnet to draw newcomers. Most of these newcomers were concentrated in the gold-mining camps of the Yukon Territory, at Juneau and at Nome, while the remaining tiny population was spread out in a territory more than twice the size of Texas with a general coastline of 6,640 miles, and almost 34,000 miles including all its islands. Following the sparsely populated coastline, the little schooner's sailing distance to Attu was to be nearly 3,000 miles.

At twenty-three years of age, Dad now had ten years under sail, most of them as a seaman on full-rigged ships. Those were the waning years of sailing vessels, lovely to look at and much romanticized today. But the life under sail was a hard one for the crews. The breed of seamen produced by the full-riggers no longer exists. They were men who could splice any line, whether wire or manila, double as acrobats handling sails in the high rigging and, perhaps most important, seamen who could tolerate a very hard existence. Those were important years for Dad because they gave him the skills and prepared him for hardships and adventures that must at times have made him long for the "luxuries" of life on the sailing ships.

A picture of Dad, taken with some shipmates at the port of San Francisco during the time he was sailing from Puget Sound, shows a burly young man with big arms developed from hauling so many sails, wearing the small-billed seaman's cap so popular still with Scandinavians and boaters around the world. At some five feet, ten inches, he was taller than average for a man of the times. Even then he wore two "trade-marks" that he would wear all his life: a bushy mustache; and a vest that he always wore both afloat and ashore, crossed by a chain from a pocket watch tucked into the vest. Not so easy to see was a mermaid tattooed on his inner forearm, an attractive figure with the full hips so admired in the late 1800's. The mermaid was tattooed all in one color, a light blue. It's likely that the tattoo was done aboard one of the sailing ships, which often carried tattoo artists among the crew who were skillful, but didn't have equipment and dyes for creating tattoos in several colors.

Dad and the Johnsons sailed from Puget Sound in their small sailboat sometime in the early spring of 1897, following the inside passage through British Columbia to the panhandle of Alaska, known today as Southeast Alaska. It was there, at Port Gardner on Admiralty Island, that Mrs. Johnson gave birth to a daughter who would know no other home than that 42-foot schooner for the first several years of her life. The baby was brought into the world on the boat, with mother and daughter attended to only by the two men aboard.

Although Dad never talked very much about the hardships of that voyage, I can guess at many of them because I went to sea myself and served for many years in Alaskan waters. I know the 100-mile per hour winds and ice of the Bering Sea, and the treacherous navigation around the Aleutian Islands. In particular, I remember a 1945 voyage to Attu when I was captain of a Liberty-class ship carrying military supplies for the planned World War invasion of Japan. The charts furnished to us were sketchy, making navigation a constant worry. When I returned to Puget Sound and saw Dad, I mentioned my concern about navigation during the trip. He laughed, and pointed out that during his voyage on the schooner in 1897 they had for navigation only a hydrographic chart of the entire Pacific Ocean, intended mainly to show ocean currents throughout that huge region!

Somehow this little party managed to reach Attu despite the dangers, and spent the winter there trapping blue fox. Not very successfully, as it turned out, but they managed to survive and to move further east in the island chain the next summer. The winter of 1898-99 was spent at Atka, where Dad worked in the trading post and began to catch up on the education he'd missed while at sea. It was there, he said later, that he first learned his three "Rs."

Disappointed with trapping, the partners now changed their business to that of hunters of game animals, providing meat for the mining camps which dotted the coastal areas of Western Alaska. They also shot bear and sold their skins, using a .45-90 rifle. That is a very large gun, such a size being necessary to safely drop the great bear in that region. They managed to eke out an existence this way, spending the next couple of winters at Unga Island, just off the Alaska Peninsula. Dad again worked in the local trading post, there being little else to do in these outposts during the winter. Reading was one of the few things people had for recreation while isolated by the winter storms, ice and snow, but reading materials were in short supply.

While working in the trading store at Unga during the winter of 1901-02, the storekeeper complained about the lack of reading material and asked Dad if he had anything to read that he might share. Dad was sleeping that winter in the trading store bunkhouse, which was so cold that he had put newspapers under his mattress to provide some insulation. Remembering that some of them were quite recent, probably from a ship calling at the port before winter set in, Dad went to the bunkhouse and pulled them from under his mattress. In one of the papers from Seattle, he and the storekeeper read about the assassination of U.S. President William McKinley, who had been shot by an anarchist terrorist while welcoming citizens at the Pan American exposition on September 6, 1901.

I haven't seen a copy of that newspaper, but I'm certain there was a lot to read in it about gold. "Gold fever" hit thousands of Americans about the time Dad first left Puget Sound for Alaska, many going to the Yukon Territory of the interior, most going through that territory bound for the Klondike region across the border in Canada. Closer to Dad's area of operation in Alaska was a strike made in 1898 at the barren, unpopulated Cape Nome, about 150 miles south of the Artie Circle on the Seward Peninsula. Unlike other strikes that required the hard work of panning interior creeks, or heavy investment in machinery for digging out veins of gold, this one involved only the dull black sand of Cape Nome's beach, laced with gold. Known as "poor man's digging" because of the relative simplicity of extracting gold, the strike brought a rush of fortune-seekers from the U.S. Within little more than a year a town of more than 2,000 people had been built, with two saloons, several bakeries and laundries, and a dozen merchandise stores.

Whether it was reading the newspapers during that winter of 1901-02 or something else, Dad was somehow attracted to that strike at Nome. In the spring of 1902, he sold his share of the schooner and departed amicably from the partnership with Mr. Johnson. Arriving at a town that had grown to 100 saloons, serving some 20,000 people, Dad staked out a claim known as "Dexter Hill," probably in the tundra behind the Nome beach itself, as the sands had by that time given up most of their gold. To help with the hard work ahead he hired a Dane from Bornholm, named John Jorgensen, and the two settled down to the miners' life at Nome. Their work during the winter consisted of thawing the frozen earth with boilers, heated by coal shipped in during the summer at outrageous prices, digging out and piling up the dirt. During the short summer, when water was available, they sluiced the dirt to extract gold.

When the ice broke up in the early summer of 1903, making shipping possible again, Dad left for Seattle to buy supplies for the mining venture. John Jorgensen suggested that Dad look up his sister in Seattle, who had arrived a year earlier and was working as a mother's helper for a family in Ballard, a part of Seattle populated mostly by Scandinavians. Dad and Emelia Jorgensen hit it off immediately, seeing each other often during Dad's visit. As family legend has it, towards the end of his six-week stay Dad turned to Emelia during one of their dates and said: "Well, I'm going back to Nome soon. Shall I buy one ticket or two?" My mother must have replied "two," for they were married in that August of 1903, and departed shortly after that for Nome.

That was a wild place for a man to take his bride. Crime and rowdiness were the rule of the day, and living conditions were hard. My parents lived in a cabin built of imported lumber on the sandspit at Nome, heated by that expensive coal. During the severe winters the snow packed up to the roof of the cabin so that a tunnel had to be shoveled out to reach the door. With so many people crowded onto a beach, sanitation in the town was a terrible problem. So bad, in fact, that public toilet facilities were eventually built and everyone was required to use them at 10 cents a time.

Their first child, my older brother George Dexter Christensen, was born in late June of the following year, given the middle name of the mine site. I was born two years later in 1906, about the time the gold mines at Nome started playing out. By 1908, my dad figured that he had extracted as much gold as was profitable out of his claim, so he and my mother, my brother and I, packed up and left Alaska on the steamship *Ohio* in the spring of that year. One of my parents' last acts before leaving was to get birth certificates for George and me. He and I have gone through life sharing one piece of paper certifying that Niels and Emelia Christensen had two boys "born on the sandspit at Nome," giving our names and dates of birth.

Dad's love of Bainbridge Island had stayed with him throughout the years in Alaska, so it was there the family headed after the *Ohio* docked in Seattle. The gold claim at Nome had not produced a huge fortune, but it did give my parents enough money to buy a piece of land at the head of Eagle Harbor, a quiet and pretty little cove; enough to build a home on the land; and enough left over from that for Dad to form another partnership. This one was with a man who had built a 50-foot work boat, then had run out of money to buy an engine. Dad bought the engine, they christened the boat *La Blanca*, and went into business on Puget Sound. The business was a combination of towing, hauling berries from the farms on Bainbridge and nearby islands into Seattle, and hauling dynamite and blasting powder from the Du Pont plant near Tacoma to the many logging operations around Puget Sound. Within a year, Dad bought out his partner and ran the business by himself.

The third child in the family, my sister, Hannah, had actually been on the trip aboard the *Ohio* from Nome to Seattle, as my mother was pregnant then and gave birth to Hannah shortly after arrival on Bainbridge. Hannah was followed by two more boys, Paul and Nels, and the last child born to my parents was a daughter they named Ellen.

Those were happy days! Dad enjoyed his work. He put us boys to work along with him on *La Blanca*, acting as deck hands, helping load and off-load berries, learning seamanship and coming to share Dad's love of the sea. Mama also enjoyed her life on the island. Raised at the town of Hassle on Bornholm, she had a strong religious background in the Baptist church. But she had to content herself with a Congregational church in the early years of life on Bainbridge, as the island had no Baptist church. Most of the children went through a Sunday school in the Congregational church that still stands in Winslow. Later, though, a Christian Alliance church came to the island with a religion more fundamental than that of the Congregationalists. Finding this closer to her Baptist upbringing, Mama became a founding member of this church on Bainbridge.

Her home life was devoted to keeping a husband and six children fed and clothed. She raised a garden, and especially enjoyed raising animals. These included many chickens, a pig raised each year and slaughtered in the fall, some sheep, and after some years the long-held dream of owning a cow. She considered her animals a hobby as well as a provision of food for the table, an interest which probably came from her early years on a Bornholm farm before moving to Hassle. There was no refrigeration in our home, so all the vegetables she raised had to be canned to preserve them for wintertime use, and much of the meat was salted.

It was a quiet and content life we led as a family there on the island, so calm and uneventful compared to Dad's early adventures at sea and in Alaska, and Mama's own share of those adventures at Nome.

CHAPTER TWO
Early Days On Bainbridge Island

THE *FLORENCE K.*

MY OWN EARLIEST MEMORY of those days comes after that of my older brother, George, who was four years old when we arrived in Seattle from Nome aboard the steamship *Ohio*. George doesn't recall anything about Nome, or about the trip south, nor does he remember what must have been a pretty exciting arrival in the bustling seaport of Seattle. What he does remember is landing on Bainbridge Island, craning his neck to look up at some monstrous green things, and being told they were trees. Born in Nome and raised to that age there, George had never before seen a tree close up.

We had come to an island rich in history, though a very brief one. The first European discoverer was Captain Vancouver, who anchored at the tip of the island in 1792 and left thinking he had been on a peninsula. It was not until 1841 that an American explorer, Captain Charles Wilkes, discovered this was an island and named it after Captain William Bainbridge, an American naval hero of the War of 1812. The island was settled and grew fast after that, with most jobs coming from the timber industry. Lumber mills were built at Port Madison and Port Blakely, the latter quickly becoming the largest lumber mill in the world. It was that mill which had attracted the many sailing ships from around the world that my father had seen when first visiting there in the early 1890's. Shipyards had also been built in both communities. The Hall Brothers Shipyard, which became famous for its lumber schooners built at Port Blakely, had moved to Winslow five years before we arrived. There were upwards of 1500 men working shifts around the clock at the Port Blakely mill, and several hundred people lived in the Winslow area, where we had settled.

I don't have any memory earlier than the Alaska-Yukon Exposition in Seattle in 1909, when I was three years old, and that memory centers on a single incident. My parents and we children undoubtedly crossed Puget Sound from Bainbridge Island on one of the "mosquito fleet" boats that carried passengers and freight between Seattle and many small landings around the Sound, probably then going by streetcar to the fairgrounds on what is now the University of Washington campus. But I don't recall any of that. What I remember is standing with my mother at a stall on the fairgrounds, where she had just bought a new-fangled item intended for peeling potatoes. Why I remember that I can't say, but probably it is because my mother was so proud of that potato peeler. It kicked around in the kitchen utility drawer at home for years.

My next earliest memory is of the mosquito fleet boats, particularly the *Florence K.*, a 250-passenger steam vessel that called on a regular schedule at our home town of Winslow. There were several men who owned her in partnership. One was Ebeneezer Franks, who served as chief engineer. Two others were both sea captains who were in the partnership but didn't run the boat. Captain Jensen was one of them, a Dane from my mother's home island of Bornholm, who used to call often at our home. These boats meant a lot to us because as children we loved the trips to the "big city" of Seattle.

Mama, on the other hand, was a lot less enthusiastic about going to Seattle. In her case it wasn't the city that put her off, but the fact that she was afraid of the water. Whenever it was time to buy clothes or other items not available on the island, she would send us up a knoll between our home and Eagle Harbor, where the boat landings were located, asking us to tell her if there were any whitecaps out on the Sound. I think her fear of the water came from having traveled from Denmark to the United States by boat, and then sailing to Nome and back, which must have been pretty rough trips. But we liked to go on the boats so much that we always reported a calm sea, no matter what the weather. Once on the boat we would run and scream around on the deck, and Mama would haul us back into the cabin with a warning that we'd fall overboard if we kept up that kind of nonsense.

One trip made regularly on the *Florence K.* was just before school started in the fall, when my father and mother would take all of us to Seattle to buy school clothes. We shopped at Schaeffer Brothers Clothing, located at 2nd and University, where Dad would buy each of us a new suit. For the boys it was always knee-length wool pants buckled just below the knee, and a matching suit coat. While these were supposedly our school clothes, to school we actually wore the outfit from the year before and saved the new outfit for church and other "Sunday-go-to-meeting" affairs. I was twelve years old before I got a pair of long pants, a pair of corduroys that I bought in Blaine, up near the Canadian border, with my own earnings from working on Dad's boat.

The Seattle waterfront was an exciting place for youngsters to be in those days. We went in with the boat at old Pier 3, where Ivar's Seafood Restaurant is located now, alongside the waterfront fire station that is still there today. The pier and surrounding streets were busy with freight being hauled to be put aboard the boats, the hauling being done by three-horse drays. They would be carrying hay, feed, and other freight, items not available on the islands, which were loaded onto a freight deck on the boats. My mother and dad were constantly warning us to stay out of the way of the wagons. Watching that activity was just part of the excitement of making the all-day trips to Seattle.

Our home life was a happy one. We fought like all children do, but generally got along pretty well. My older brother and I would fight quite a bit, with George usually getting the better of me. He was left-handed and somehow I just couldn't get around that southpaw. Outside of that we played and did our chores together with no problem.

The house my folks had built when they returned from Nome had no running water or electricity, so there were plenty of chores for us. Water was pumped by hand from a well on the property. Kerosene lamps lighted the house until 1915, when electric service was brought in and the house was wired for lights. That brought another blessing in the form of an electric pump on the well, allowing Dad to install indoor plumbing. Up to then we used an outdoor toilet, a "Chick Sales" as they were called. The hole would fill up after awhile, when we'd have the chore of digging a new one nearby and moving the building over it.

Another chore I remember was chopping kindling, something I was supposed to do every evening. One night I played around the yard so late that I forgot to cut the kindling before going to bed. It was cold the next morning when Dad got up to build the fire. When he saw there was no kindling he hauled me out of bed and waltzed me down to the woodshed in my nightshirt. The first thing he made me do was to cut a piece of kindling for paddling my fanny. Then I had to cut the rest of the kindling for the day. It was a job I never again forgot to do!

Still another chore was cleaning out my mother's chicken house. After cleaning it we had to wipe kerosene on the chicken roosts to avoid the little red nits that infested the birds. Mama did enjoy her chickens! She always had at least three or four dozen, selling the eggs not used at home to the local grocery store. She was always bragging about how much money she made from her chickens, which was a joke in the family because Dad bought all the feed and other necessities to keep her "business" going.

Some of the Scandinavian immigrants tried to keep up their language in the family. Mama belonged to that school, speaking Danish most of the time around home, saying Danish prayers with the children and always reminding us of our Danish heritage. Dad, on the other hand, was very firm in seeing to it that his children were raised as Americans. He and my mother had one of their rare arguments over use of the Danish language in their home, with Dad wanting only English spoken so that the children would be prepared to live and work in English-speaking America. The argument was settled with a compromise. Mama promised to speak only English at home if Dad would promise to stop chewing tobacco and spitting the juice into cans. That settled, Dad took up smoking a pipe, which became another of his "trademarks," clenched in his jaw both afloat and ashore. From the standpoint of language, that left the three of us older children with a smattering of Danish, and the three youngest with almost no knowledge of the language.

We almost always had hired help around the house. Mama had rheumatism pretty bad during her childbearing days, making her hands awfully painful. Since Dad made a good living with his boat business, they could afford to hire a Japanese lady who came each Monday to wash all the clothes, and again on Tuesday to do the ironing and put them away. We also had high school girls through the years who came in the afternoons for housekeeping. They earned five dollars a week, good wages for those days. I remember one especially, Annie Tobey, a jolly girl who lived across the bay from us in Eagledale. She would row across and tie up her boat at our float to come to work. At the end of one week Dad gave her a five-dollar gold piece for her wages, and we walked with her down to the dock. Just as she was getting in the boat she dropped the gold piece overboard, where it was hopelessly lost in the mud. "Oh, well," said Annie, "I'll get another one next payday."

There was a lot for children to do around our place. We had a float and dock right in front of our property with Dad's boat, *La Blanca*, moored out in the bay. There were always three or four rowboats around. From a young age we did lots of fishing in those boats. Mostly we went bottom fishing in the harbor, catching a lot of tom cod. These are small cod that look almost like a trout, which Mama wouldn't eat because she thought they were wormy. She didn't mind frying them up for us, though. If she was too busy to do that after we caught a mess of fish, we'd take them over and give them to Mrs. "Blacksmith Johnsen," the Norwegian wife of the local blacksmith. She always appreciated getting our fish. There was no welfare in those days, or social security, so neighbors often gave or exchanged things, helping support those who were in a bad way.

By the time I was ten or twelve years old I was permitted to row outside the harbor with my next youngest brother, Paul, to fish for salmon. We'd row out into Puget Sound off Wing Point, trolling a line with just a simple brass spoon on it. It wouldn't take more than a couple of hours to catch a half-dozen small blackmouth salmon, a treat even Mama enjoyed.

We spent a lot of time either on or in the water. Swimming was very popular. It seems we spent most of the summers in our swimsuits, playing around the dock or swimming out to *La Blanca* when she was in port, to dive off her. Lots of friends joined us in the swimming and other play, which included building a shack in the woods where we played "pirate." We scrounged the lumber for that in different places, including the bay. Besides the occasional piece of lumber, the bay often was full of log rafts being held there for the lumbermills, huge first-growth logs that were four and five feet in diameter. We'd run on those, scaring my poor mother out of her wits, and also used to go down to the shipyard at Winslow to run all over the ships under construction and in for repairs.

Mama had good reason to be worried about us playing around the water. One late summer afternoon when George was seven, I was five and Hannah three years old, we were picking blackberries from bushes on a low bank over the water in front of our home. Hannah slipped, and with the tide just past its high point went right into deep water. George and I had enough sense to run up to the house to tell our mother what had happened, but by the time we got back Hannah had drifted 30 or 40 feet out into the bay, and was moving out with the tide. Two things saved her. One was the clothing little girls wore in those days, a white dress with lots of petticoats that both floated her and made her easy to see. The other was that some girls were camping nearby. A couple of them saw Hannah floating by, jumped in a canoe and rescued her. After getting ashore they gave her a sip of brandy, then brought her over to our place to put her in Mama's arms. My Uncle John rowed across the bay to get Dr. Kellum, our family physician, who found my little sister in good condition considering the experience.

The next summer I gave my mother a terrible scare by running up to the house to tell her "Hannah's drowning herself again!" Hannah was actually just picking blackberries, which my young mind had interpreted as being the way one drowned.

The tidelands in front of our property were full of clams. It was one of the best clamming beds in the area, drawing among others some Indians from the Suquamish and Port Madison tribes. I can remember playing in our yard when a number of dugout canoes would come gliding around the point, the women paddling and men steering. They would pull right up in front of our place, and both men and women would dig many sackfuls of clams. I don't know why, but for some reason we never talked with the Indians. It may have been that we were a little frightened of them, as it hadn't been very long before that when the last of the wars with the Indians took place in eastern Washington.

I can still remember my first day at school. I had a loose tooth which my mother's brother, Uncle John, pulled out just before I left for that first day. I made a great fuss over it. "That's a good sign," Uncle John said, "Every good boy gets a tooth pulled before going to school." We milled around in the Winslow school yard, meeting new friends, before assembling to march to our classes. The routine was that boys and girls marched together, two by two, which I didn't like very much because I had to march with a girl who was taller.

It didn't take me very long to get into trouble. Below the school was a pond, where we were forbidden to go. A couple of other kids and I went right down there to play in the water. I got my brand new hightop boots all wet. Our teacher sent us to the principal, a big husky woman who gave us a ruler across our stern ends. It was a lesson I didn't forget.

The only high school on Bainbridge Island was included with our elementary school, so we had all twelve grades there. The first eight grades were on the ground floor of the school, with high school on the second. Most everybody went to that high school, though a few took the boats to attend Queen Anne or Broadway high schools in Seattle. On the first floor there were two grades in each classroom, about two dozen children making up both grades. It seems to me that the teachers were more dedicated in those days than now. I remember one in particular, Miss Athens, a lovely older lady who was my third grade teacher. She was my favorite, a teacher really dedicated and interested in her students.

The lesson after getting wet in the pond held pretty well with me until fourth grade, when I again got into trouble. There was a boy who sat in front of me in class, Burt Case, who bothered me all the time by turning around to copy from my work. I told Dad about it, who advised "Why don't you punch him in the nose?" I did. That night, Burt and his brothers laid wait for me as I walked to the grocery store on an errand for Mama. Just as I was in the middle of the bridge on the way to the store, Burt and his brothers lined up, some at each end of the bridge. They caught me there and beat the hell out of me. Fortunately Mrs. Nakata, the wife of the Japanese barber, saw what was happening and came out with a broom, chasing the Case boys off and taking me into her husband's shop to clean me up. She shook her head as she worked on me, saying "Awfury bad boys, awfury bad boys." Dad wasn't very sorry about it when I got home. He never openly pitied us much, as he felt we had to learn to defend ourselves.

We had some fun at school and afterwards, too. Baseball was the big sport then, a game we played almost every afternoon before having to go home to do our chores. It wasn't organized ball, just sandlot games. I used to hide my books in an old stump near the field during the game, often forgetting to pick them up before leaving for home. "Where are your books?" Dad would ask. I'd tell him that I left them at school, but that story stopped working after a time. So I had to remember from then on to pick up my books after the game each day, ending the education of that poor old stump.

In the winter of 1916 there was a terrific snowstorm on Puget Sound, with about four feet of snow gathering on the ground around Winslow and staying there for quite a long time. There were no "snow days" then to free us from classes, so we continued walking to school every day. Dad would tie burlap bags around our shoes in the morning to keep our feet dry, and he and Mama insisted that we come home through the snow for lunch each day, a hot meal to keep us going through the cold.

As I got older I became good friends with a fellow named Clinton Cave. Clint was about three years older and had a car, a 1913 Cadillac. We used to have a lot of fun taking trips in that car. There was a ferry running then from Point White, at the southwest corner of Bainbridge Island, across the narrows to Waterman. It was just a small boat pulling a scow that carried about four cars, the only ferry service the island had except for my father sometimes hauling cars to Seattle on a scow pulled by *La Blanca*. Five or six of us would pool our money and take Clint's car across on that ferry, going up Hood Canal to Liliwap Falls. We made quite a few trips like that when I was in my early teens.

Another time I got in trouble was when Clint and I were horsing around one day in Winslow, throwing rocks at each other as boys will do. I let one fly at Clint while he was standing in front of the post office, he ducked, and the rock went through the window. I didn't know what to do, other than haul out of there as quick as I could. A couple of days afterwards I was helping Dad load *La Blanca* when a telephone call came for him.

"Holger, come here!" he yelled, coming back to the boat. "Did you break the window up there in the post office?"

"Yes, I did," I told him, starting to explain how it had happened. He interrupted, saying it didn't matter how it was broken and asking again if I had done it. I said I had.

"Well," Dad said, "Charlie Morrell called up to tell me that you broke his window. He wants two dollars and fifty cents to have a new one put in." Charlie was the owner of the building. Dad reached into his wallet for the money, telling me to take it up to Charlie. "We don't want anybody stealing the mail," he said.

That was typical of my father. He was a strict disciplinarian, but he had a sense of humor and fairness. Dad knew I wouldn't break a window like that on purpose, so he went easy on me.

Life continued happily at home and school up to my sophomore year. Although I didn't have any serious trouble, I was a rebel of sorts and had a run-in with the school over a science class. A couple of friends and I were fooling around with an experiment using acids, mixed up the wrong stuff and ended up with a broken test tube and the classroom full of smoke The teacher kicked us out of the class. I was mad because I had liked that class, so I just quit school. Typically, Dad said: "If you don't work, you don't lie around here." It wasn't very long before I had a job as deckhand on the old steamer *Bainbridge*, running between Winslow and Seattle.

CHAPTER THREE
La Blanca

LA BLANCA AT ANCHOR

IN THE EVENINGS Mama would huddle all of us together in the living room of our home, to say prayers in Danish. When Dad was off on his boat, working in some part of the Sound for a week or more, the prayers were often for him and his crew. Mama always worried about Dad being out on *La Blanca*. Hating the sea, having no interest in it, she had no way of knowing that Dad was a marvelous and safe seaman. So she worried and worried, and when we boys were older she shared those worries with us.

We, of course, loved the boat and the sea. Though our mother might dislike the idea, *La Blanca* was the center both of our family universe and of community activities as well.

She was a comfortable boat, about fifty feet overall, with accommodations in the focsle down forward. There were two permanent bunks there, with a bench on the side that could be made up into a bunk. A nice little galley stove, fired by coal, used to cook meals and as a heater. There was also a bunk in the wheelhouse where Dad slept. I have memories of many cold nights while crewing for Dad, sleeping in the cozy focsle, warmed by the stove, after a good meal.

The engine Dad bought for *La Blanca* when going into the original partnership was a four-cylinder Eastern Standard. It was a long-stroke engine that burned "distillate," a very low grade of gasoline, firing the cylinders with a make-and-break system that had no sparkplugs. It was a good, faithful old engine, and though it took a bit of nursing it lasted until 1926 before having to be replaced. *La Blanca* was always well maintained, with Dad doing most of the work on her. Every spring it was time to paint the bottom, and unless there was some other maintenance necessary on the underwater parts of the boat Dad didn't put her up on a marine way. Instead, he'd find a nice gravel beach when the tides were right, running her up on the beach at the first high tide so that she'd be high and dry when the tide went out. Then he'd run a line around a tree ashore, back to the boat where a block and tackle were made fast to the mast. As the tide went out he took up on the line, leaning the boat in towards shore so one side of the bottom would be fully exposed when the tide went out. This way he had time to scrub and copper paint the bottom during one change of the tide. At the next high tide he would turn the boat around, going through the same process to paint the other side. It was only when there was some mechanical work to do, like replacing a stern bearing or working on the rudder, that he would go to the expense of putting *La Blanca* in the shipyard.

She was a good sea boat, able to take any kind of weather. Dad made one open ocean trip down to the Hoh River on the coast, for a man named Charlie. Charlie owned a trading post there, which he supplied with goods from Seattle carried on his own boat. On this occasion he had wrecked his own boat going in the river entrance, losing all the goods he was carrying. So Charlie hired Dad to carry new supplies on *La Blanca*, going along on the trip to act as pilot for the river entrance.

The weather was bad when they got there, with seas breaking over the rocks at the entrance. As they were well into it, with Charlie steering, Dad saw a wave surging over a rock right alongside the boat. "I'll take the wheel, Charlie," said Dad, "I thought you said you could pilot me in here." "Well I can," Charlie replied, "but don't you worry. The boat is new and the rocks are old." Somehow they managed to get in there safely, though they were stormbound for a week with my mother terribly worried about what had happened to Dad.

Dad started taking us boys out on the boat while we were still quite young, tethering us to the mast with a line around our waists just long enough to keep us from going overboard. We enjoyed it immensely. As we grew older we started crewing for him. He was a fine instructor in seamanship.

Many memories of my early days on *La Blanca* center around community use of the boat. There was no hospital on Bainbridge Island in those days, so Dad was often asked to provide ambulance service from the island to Seattle, a trip that took about an hour. That old boat carried many a woman in labor, and many an injured man for hospitalization in the city. Oftentimes the people were unable to pay for the service, but Dad never hounded them for the money. It was just another part of the spirit of being neighborly.

One trip on which I crewed was to take a man who had almost lost his leg when he was hit by a piece of flying steel in the Winslow shipyard. Another was a man helping to dismantle the Port Blakely sawmill, who had his arm nearly taken off when it was caught in a cogwheel. They laid him in the pilothouse bunk in terrible pain, with a doctor along to take care of him. The man was a heavy smoker, so to ease the pain I stood alongside the bunk lighting one cigarette after another in my own mouth and then putting them in his. I must have been about twelve years old at the time.

On another trip, Dad picked up a Japanese boy at Port Blakely who had been shot in the head. Somehow the bullet hit the skull and then just curved around his head under the flesh, so he lived.

It wasn't always ambulance service, though. There weren't many roads on the island in those days, so many communities could communicate with each other only by boat or by horseback on trails. A very old friend of Dad's, Captain Oliver, died near our place at the head of Eagle Harbor and was to be buried at Port Madison, which couldn't be reached by road. So Dad transported the body on *La Blanca*.

On another occasion, he was commandeered by the deputy sheriff in Winslow to give chase to two robbers who were escaping by rowboat. He, Uncle John, and the deputy got aboard the boat and took off down the harbor after the robbers, with the deputy up on the bow with a shotgun. "Halt, in the name of the law!" the deputy shouted at the poor, scared robbers. Uncle John, who had been in the Danish Army and knew how dangerous guns are, went up to the deputy. "Here, Fred," said Uncle John, "let me take that gun from you. You're liable to shoot somebody with it." I don't know what happened to those robbers, but at least they weren't shot.

We had some fun with that boat, too. On a nice summer day when *La Blanca* wasn't chartered out, such as on the Fourth of July, my folks would invite a group of friends to come along with the family on a picnic. Dad would run us to a bay with a sandy beach, anchoring out and ferrying the party ashore by skiff for some wonderful food and good times. I especially remember a glorious day when a bunch of friends came along with us to Bremerton for a picnic and to see a fair being held in the city. Those were times we children enjoyed very much.

One of my father's regular contracts with *La Blanca* was to haul berries from the islands into Seattle during the early summers. Strawberries from Bainbridge were loaded at the head of Eagle Harbor, brought there by the Japanese farmers on the island, and there were so many that Dad sometimes had to charter an extra boat to handle all of them. An exciting night that lives in my memory was during the strawberry season in 1918, when George and I were crewing for Dad. There were so many berries to transport that we pulled a barge to Seattle to carry the overflow. Our usual procedure on arrival at the Bell Street wharf was to shorten the towline and bring the barge alongside before landing at the float. The berries were then offloaded onto elevators that took them up to the warehouse to be stored in different lots. By then there were many trucks in use, which picked up the berries and took them to market.

Just as we were taking in the towline on this particular night, the whole roof of Pier 11 just south of Bell Street blew off in a sheet of fire. Pier 11, which was used for wartime ammunition storage, was engulfed in flames. Dad moored the barge and boat despite the fire. While George and I stayed on the boat, Dad took the shipping manifests up to the office along with Ray Parfitt, an islander who was crewing for him at the time. While they were on their way to the office an explosive shell came roaring out of the fire across the way, just missing them and imbedding its shrapnel in one of the two-foot square posts holding up the warehouse. Dad came back to the boat and had us kids hide behind those big posts while he completed his business.

There was a large crowd of people at the armory, watching the fire at quite a distance. A boy in the crowd was killed when he was hit by one of the shells exploding out of that fire.

A big steamship, the U.S. cable ship *Burnside*, had been tied to Pier 11 when the fire started. Even if the boilers are fired it takes a steamship quite a time to warm her engines before getting underway, so she was badly scorched on one side before a big steam tug finally got hold of her, dragged her out into the stream, then brought her into the Bell Street wharf. While we stood at the wharf looking at the damage, a fire started in one of the lashings holding down gasoline tanks on the deck of the *Burnside*, used for the small boats on the ship. Ray Parfitt ran up to warn the crew, the fire was put out, and the excitement died down. That was a night we talked about for a long time.

We did a lot of business with the Japanese on Bainbridge Island. At the time my dad was hauling berries for them, into the 1930's, they had a barreling plant at the head of Eagle Harbor, where they filled 50-gallon barrels with the berries and added about a sack and a half of sugar, bringing the weight of each barrel to about 500 pounds. In the early days, we hauled those barreled berries on La Blanca to cold storage at the Spokane Street Terminal, and later to the City Ice Company in Ballard.

There were about 130 independent Japanese farmers on the island then. When we were hauling their berries it was hard for us to write down all those Japanese names, so each farmer was assigned a three-digit number. In shipping the fresh berries, number 606 might have 150 crates of berries, while 909 might have more or less. They never asked for receipts, always trusting my father to credit them with the right number of crates.

The Japanese deserve lots of credit for developing Bainbridge Island. They arrived there when the raw land was covered with stumps, clearing acres and acres of land for their farms.

They were good, honest people, wonderful to work with. We never had any problems with them, and if it hadn't been for the Japanese the island wouldn't have developed as fast as it did. It was a raw deal for the Japanese when they were interned during World War II, a sad affair. Many of them never came back to reclaim their farms at the end of the war.

The berry season lasted about six weeks. After that, my father chartered *La Blanca* to the Western Packing Company at Blaine, taking us boys along as crew. We were hauling salmon, running out to the fishing banks at Point Roberts and others around the San Juan Islands, picking up from gillnet boats there, and getting most of the fish from a scow moored by the packing company at Alex Bay. The scow served as a sort of headquarters for many of the gillnetters, who would deliver their fish there in the morning after a night's fishing. There were tallymen on the scow, who kept track of each fisherman's catch and then put the fish on our boat when we called there. We'd haul the fish into the cannery at Blaine, where sometimes we had a chance in the evenings to go up into town to see a show.

Besides a contract to carry blasting materials for Du Pont, hauling berries and working for the fish cannery, Dad had a lot of odd jobs with *La Blanca*. One of the fellows he worked for was a sail maker in Eagle Harbor named Jacobsen, also a Dane. Jake made a lot of sails for the sailing vessels in the early days, which Dad would deliver to the ships. He also dealt in all the running gear necessary for those vessels, canvas and the like, and was a junk dealer besides. Jake had a boat himself, the *Hope*, which didn't have the capacity for hauling that *La Blanca* had. He would use her for going out to visit the ships and take orders.

One time, Jake partied in Mukilteo while the *Hope* lay there alongside a sailing vessel. Jake didn't pay proper attention to the fact that a big northerly storm was blowing hard down Saratoga Passage. The *Hope* broke loose and went in under the dock, breaking off her mast, knocking the rudder loose from its fittings and leaving him in an awful fix. He finally got her secured so there wouldn't be any more damage, then called my father to tow the *Hope* back to Eagle Harbor.

I was either on school vacation or it was a weekend, because I went along on the trip. We bucked that heavy northerly wind all the way up there. It was still stormy, and night was falling when we finally got the tow underway and headed back to Eagle Harbor.

It must have been the fall of the year, as the gillnetters were out. These salmon fishermen have long nets stretching out from their boats, suspended by corks and marked at night only by a small light at the end of the gillnet and one on the boat. They're very hard to see, especially at night, and even more difficult to pick out when the sea is running as it was that night. As luck would have it, the *Hope* ran over one of those nets, which caught in the broken rudder skeg of the boat so that we were dragging the net along with us. First thing we knew, a Greek fisherman was coming after us, whistle blowing. He was very mad, and took out a big, wicked-looking knife. We were pretty worried there for a minute, but he took the knife and cut the net loose from the *Hope's* skeg, yelling "Go on, get out of here." Which we did. The *Hope* lay on the beach for quite a time. She was an awful mess, but Jake eventually got her repaired.

I remember many other trips with Dad while he was working for Jake. They were good friends. The two of them would sit in the pilot house having a good time, swapping stories of the old country, laughing and slapping each other on the leg.

We weren't quite as able as the grown men in the crews, but Dad appreciated our work and paid us for it. The end of that first summer hauling salmon, just before returning home, was when I bought my first pair of long pants in Blaine. I guess I wanted to be a man. Mama was astounded to find I had grown up so fast when we got back home.

It was time then to get back to school, back into the old routine, but my father had a hard time getting men to crew for him because of the shortage of manpower during the war. So he would take me along on shorter trips, trying not to interrupt my schooling because he placed a lot of importance on that. By that time he had a lot of faith in my ability to handle the boat, even though I was only twelve years old. He would lay down in his bunk to sleep, trusting me to navigate and, I suppose, having faith that I had enough common sense to call him if I got into trouble. I think now that he was deliberately trying to give me a sense of independence.

On one weekend trip during that period, Dad planned to get me back in time for school on Monday after hauling a load of dynamite from the Du Pont plant. A real storm blew up just as we arrived off Du Pont, making it impossible for us to get into the dock. We lay there in the bay, waiting for the storm to calm before we could load and deliver the blasting materials, making me a day late getting back to school. When the teacher asked for an excuse, Dad wrote: "Late getting back. Stormbound at Du Pont."

One November he did take me out of school for a run from Du Pont, carrying powder to a logging company at Discovery Bay. The logging company needed the powder to clear out a right of way for a railroad being built into the woods to haul the logs out.

There was another lad along on the trip, which turned out to be pretty rough. Heavy seas came up as we approached Point Wilson, and the engine was acting up. Dad left Warren and me in the wheelhouse to take care of the navigation while he stayed in the engine room to nurse along that big gas engine. It was another example of his faith in me to get through rough weather while he kept that old propeller going around.

It was a chilly, clear evening when we got to the logging camp pier. The rail line extended out onto the pier where we would be loading the powder onto flat cars. Dad went ashore to notify company officials that we had arrived with the powder, while I started cooking our evening meal in the galley down forward.

It was just getting dark when an old logger came rowing up to the dock, back from having bought a newspaper at the company store. He and another logger shared a shack up on the beach, where there was a camp for many of the men who worked there. By now the wind had died down. Through the still air I heard the man in the rowboat yell up to his partner: "Hey, you old son of a bitch, the war is over!" It was November 1918.

I never go by Discovery Bay without remembering that night, and the many, many happy times during my days with Dad on *La Blanca*.

CHAPTER FOUR
Working on Puget Sound

THE *BAINBRIDGE*

DAD WAS THE MAINSTAY in giving us work to do, sometimes even after we were grown, but our work wasn't limited to crewing on *La Blanca*.

One of the first jobs I remember was when George and I were still little shavers. Dad had gone to an auction someplace, where he spotted some cases of patented five-gallon cans, real nice ones that had pumps on them. Since we didn't have any electricity then and used kerosene lamps, he thought they would be good for handling the kerosene. So he bid on what he thought would be a case of these cans. To his surprise, he got the whole lot of ten cases or so, packed six to a case.

Everybody used a lot of kerosene then, which gave Dad the idea that he might be able to get rid of them. He told George and me that if we could get one dollar for each can, we could have two bits out of the dollar. We each took two of these bright red cans and went out peddling them. Every place we went we peddled these cans for a dollar, so it wasn't long before we got rid of them all. We did pretty well, and so did my dad.

As young fellows around Winslow we were fortunate that the shipyard usually stayed busy. Sometimes they would call to offer us work, or if Dad had nothing for us to do after the early summer berry season we'd go down there and could quite often get a job. Usually it was as a machinist's helper for which they'd give us 52 cents an hour, not bad pay for those days. We were happy to get it, since it gave us the opportunity to buy clothes and other things, and it kept us busy and out of mischief for a week or ten days at a time. The shipyard was very good to the kids around Winslow.

In January of 1922 there was a sailing vessel, a four-masted barkentine named the *James Tuft*, that left Port Alberni on Vancouver Island with a load of lumber bound for Callao, Peru. After a day or so out of port she hit a terrific storm, causing her to get water-logged. They usually put a big deck load on those vessels, about seventeen feet high, so when this wooden vessel got water-logged she didn't sink due to the buoyancy of the lumber. The captain got her into the lee of James Island, where he dropped both anchors and managed to ride out the storm. Eventually someone saw their distress signals and sent out a big tug, the *Sea Monarch*. She was a sad looking ship when the *Sea Monarch* towed her into Winslow, one rail under water with decks and cabins all awash. The crew had to live on top of the deck load while they were being rescued. After the ship came in they unloaded the lumber on barges, on the docks and anywhere else they could put it around the shipyard, in order to get ready for repairs.

The skipper, Captain Friberg, was a friend of my father, and it was through that friendship that I got a job helping out on the *James Tuft*. I think it was because my Dad had given Captain Friberg's son, August, some summer work on *La Blanca* that he thought of me. After they got some of the deck load off I went to work for the second mate, Mr. Bland, a real old sailor man. He was seventy-five years old at the time, born around 1847, and had some pretty fantastic sea stories to tell. A tall, skinny fellow with a flat nose, probably from a lot of bar-room fights, he had a huge voice. He was always yelling at me: "Christensen, all of yez come forward." Since I was the only one with him on the ship, I'd go up to give him a hand.

My main job was scraping down the bulkheads and deck, as water had ruined the paint while she was awash. I scraped and polished the whole afterdeck. To relieve me of the monotony of scraping, Mr. Bland would give me other jobs once in awhile. He was a rough old fellow but I got so I liked him, and I think he even liked me. One of those odd jobs was to clean out the forepeak, where the salted meat and fish had been stored for the voyage to Peru. After the ship was waterlogged, and then pumped out, all the kegs had broken and the meat and fish had rotted, causing an awful smell.

We hauled all that stuff up in a bucket with a rope on it, through a hatch about four feet square with a coaming of about two feet. There was a ladder down there, with the only light coming from the hatch. Mr. Bland said, "I'll go down there to load the bucket, Christensen, and you haul the stuff up." So I'd haul it up and dump it overboard. It was pretty rank stuff. One time I was lowering the bucket when the handle hung up on a nail on a stanchion. As I kept lowering the bucket the weight of the rope broke it loose so that it fell right on Mr. Bland's head. There was a real storm brewing then. He called me everything under the sun, but it was so funny to me that I fell down laughing alongside the coaming. Pretty soon the longshoremen near the boat heard him and saw me laughing, and they joined in the laughter. It was funny to us, but I guess it wasn't so funny to Mr. Bland.

That man had a way of turning the air blue when he was mad! Another time he and I were standing aft under the poop deck while they were still discharging. Lumber had been stored there as it was every place on the ship. In order to get a full load for those ships they jammed the lumber in every cranny they could, even up under the beams. While we were standing there, the ship took a sudden list. Mr. Bland and I heard a terrible rumble that we thought was coming from the forward part of the ship, but the first thing we knew a whole tier of lumber came over on us. We were buried in it but somehow weren't hurt, just scratched up a little. That old gentleman certainly used his command of the language then, even though he wasn't hurt.

It took a month or six weeks to finish work on the *James Tuft*. After she was fully discharged they put her in the dry dock at the shipyard, re-caulking and driving wedges in the double planking inside of the hull. When that was finished and she was seaworthy again my job was finished. I then went back to work as a deckhand on the *Bainbridge*.

I'd decked on her before on temporary jobs. This time I got a steady job, making me feel pretty important. They carried three deckhands. My job was to help the mate with the gangway for passengers, which we carried with us and slid ashore at each stop. After a time one of the other deckhands quit, so I was promoted to throwing the heaving line. That's a small line with a "monkeyfist" at one end, a ball made of rope which helps the line carry when it's thrown. I'd tie the other end of the heaving line to a spring line, a large rope that would be dragged onto the dock and made fast to hold the *Bainbridge* alongside while loading.

We made six stops in Eagle Harbor. From home port in Winslow we'd go to Eagledale, cross back to the Winslow shipyard, then to Holly Landing, Wing Point Landing, and finally across again to Creosote, the last stop before heading to Seattle. One time it was raining and blowing hard as we went in to land at the shipyard, making it hard for the skipper to bring the vessel close to the dock. Since we were quite a distance away, I had to throw the heaving line hard to get it onto the dock. There were several passengers waiting there in the rain, including a lady in fancy clothes with a beautiful hat. As luck would have it, the heaving line went right towards her, hitting her in the neck and taking a turn right around her. The line was wet, dirty and covered with tar from pilings on the docks, so it's easy to imagine the mess it made of her fancy clothes. I felt pretty bad about it.

After we cleared the Creosote dock and headed for Seattle, the skipper hollered from the wheelhouse: "Get Christy up here!"

I was sure I had lost my job, but when I went into the wheelhouse the skipper was all smiles. He reached into his coat pocket, pulling out a cigar and handing it to me. "That," he said, "is the best thing that ever happened, especially to that woman." She was a haughty type who gave everybody a bad time. My job was saved.

The skipper was Captain Welfare, a great fellow with a fine sense of humor. One day I was soogying – washing down the paintwork to get it ready for painting. I was lying on the deck under a lifeboat, soogying the bottom, while the ship was rolling a lot due to bad weather. The skipper saw me there as he walked by on deck, and said: "Christy, you'd better move inboard. I don't want you falling overboard and losing that soogy rag."

After leaving the Creosote landing, I often went to the wheelhouse to steer the course to Seattle. Captain Welfare would tell stories about his early days at sea. He had been mate and skipper on the old steam tugs that used to go out to Cape Flattery to tow sailing ships in through the Strait of Juan de Fuca and down Puget Sound. I remember one of those was about a time they were inbound just off Dungeness, when a corpse was seen floating in the sea. They stopped to pick up the body, and after getting it on deck Captain Welfare and the engineer checked the pockets for identification. In the man's wallet they found $27,000 in cash. Captain Welfare told the engineer: "I hate like hell to rob the dead, but nobody'll know the difference if you and I keep this money." The engineer wouldn't have anything to do with it, though, as that money was too spooky for him.

One good thing about that job was the overtime we got on special trips late at night from Eagledale. There was a dance hall there, called Seden and Dunlap, that drew a pretty rough crowd from Seattle. They were so bad that the skipper didn't want them mixing with the regular passengers, so after our last regular run we'd go back to Eagledale to pick up that bunch from the dance hall. Those were prohibition days, the thing to drink being moonshine for the most part. There must have been lots of it at the dances, because that crowd was always drunk, unruly, and obnoxious. There were constant fights and all kinds of lewd conduct. After a time the popularity of the dance hall petered out, ending the overtime pay.

Even as a youngster I was a saving sort of person. That overtime had helped boost my bank account to the point where I figured it was time to buy a car. I was fifteen years old at the time. I rode into Seattle on *La Blanca* with Dad and a crewman from the island who was going to buy a new car. We went up to William O. McKay's on Western Avenue, and while the other fellow was looking at new cars I looked at used ones. The salesman told me a fellow who worked across the street wanted a new car if he could sell his 1917 Ford. He wanted one-hundred dollars for it. We went over to look at the car, took it out for a ride, and I decided to buy it. I put five dollars down, with the promise to come back in a few days with the balance. Next time we came in to Seattle I brought along my bank book, took a hundred dollars out of the bank, then went to pick up the car with Dad and a fellow named Humphery. He worked for the R.D. Bodle Company and could drive in the city, something I didn't want to try yet with so little experience behind the wheel.

The first thing we had to do after picking up the car was to go down to the courthouse to change the title. Humphery drove us down Second Avenue to get to the courthouse, speeding down the street so fast that I was worried. I looked at the speedometer, which showed we were going seventeen or eighteen miles an hour. "You're speeding," I told him, "look at the signs. It's a fifteen mile-an-hour limit in here."

"That don't matter," Humphery said. He kept speeding all the way to the courthouse, where I took out a title in my name and got a driver's license. I think I paid fifty cents for the license. There was no written or driving test at that time.

A ferry service had been started to Port Blakely. Humphery got me down there and safe on to the ferry. From there I would be on my own. When we got to Port Blakely I cranked the car to start it, got off the ferry all right and started up the hill. It stopped about half way up. Lucky for me I had picked up a hitchhiker on the ferry, a fellow living in Eagledale who knew something about cars. He got out, lifted the hood and discovered that the distributor cap had popped off as we drove on that rough country road. He put it back and we got home safely.

I had a lot of fun with that car for the five years I kept it. It was a Model "T" touring car with a cloth top. First thing I did was buy side curtains to keep out the weather. I ran all over the island with it, never having any serious problems but lots of experiences typical of those old cars. One inconvenience came if the gas tank was half empty and you had to climb a hill. The only way to get up the hill in that case was to climb it backwards, so the gas could run down into the carburetor.

I got laid off from my job on the *Bainbridge* once in awhile even though that supposedly was a steady job. To save money they would cut back to two deckhands, putting me on the beach. One of those times was August of 1921. Because *La Blanca* was in for repair, Dad had chartered another boat to haul berries to Olympia and put me on the run as a sort of purser, responsible for keeping track of the berries. The boat was a little steamer out of Tacoma with its own crew, but I would take a watch at the wheel sometimes. There were loganberries grown on Bainbridge Island in the late summer and early fall. We'd pick up a load of them at Winslow, stop at Vashon Island for another load, then go to Olympia to discharge the berries, making one round trip each day.

One time I was at the wheel while we were northbound from Olympia, coming down through the Narrows off Point Defiance. I spotted something floating in the water and told the captain, whose name was Olsen, suggesting that we go take a look at it. He said it was nothing, but I got the glasses and took a look at it. When I saw that it was a rowboat floating upside down, he agreed to go take a look. It was a skiff turned over, with a man laying across it with his face in the water. We got him up on deck and got all the water out of him, finally getting him to come to. That stuff we got out of him smelled awful. When he did wake up I thought he was going into convulsions. He stretched his hands out, his eyeballs were rolling around, and then this fellow who had looked dead started singing "I'm sailing back to Norway."

After we got his boat righted, we found a half-full gallon of moonshine, and a lady's sweater. He apparently had been out there drinking, had got drunk and rolled the boat over. We took him into Point Defiance dock and called the police, figuring he needed to be taken to the hospital. When they came we tried to give the bottle of moonshine to one of the policemen, but he said: "No, you'd better take this and dump it. The poor guy's got trouble enough now." We never did find out if he came out all right, or if there had been a lady with him.

My life continued like that for more than a year, sometimes decking on the *Bainbridge*, sometimes crewing for Dad on *La Blanca* or other boats he chartered, and occasionally working in the shipyard. It was a good life, one that was increasingly giving me an interest in a seafaring career beyond the bounds of Puget Sound. The time had come to start looking for a job that would take me out on the deep sea. ⚓

CHAPTER FIVE

Going Offshore

GETTING A JOB ON A DEEP-WATER VESSEL when I was young was in some ways similar to finding a berth today, and in some ways very different.

The main similarity was that friends in the trade were a great help. It was possible then to find a shipboard job by simply visiting the many vessels in ports such as Seattle, but it was much easier if one had friends aboard, or some sort of connection with the person doing the hiring — usually the skipper or first mate.

The difference was in union influence on hiring. My dad had joined the Sailors Union of the Pacific in the early 1890's, remaining a believer in the union most of his life, but that union still had little power over hiring practices when I first thought about going offshore as a sailor. Their main concern in those early days was with shipboard conditions, which were still very poor on some of the ocean-going ships in the 1920's. I played a small part in the first major seaman's strike of 1934, a strike that succeeded in gaining some union influence on hiring practices, but that's jumping ahead of my story.

In March of 1923 I was back to decking on the *Bainbridge*, after spending most of the previous summer helping out my father on *La Blanca*. At the age of sixteen years, with the experience on those boats and no desire to go back to school, I began to think seriously about a career as a seaman. It wasn't long before my first chance to sail offshore came along, in the form of a vessel brought into the Winslow shipyard to be outfitted.

The *Caesar* had been built in 1898 for the U.S. Navy as a freighter to haul military supplies. About 250 feet long, she was fitted with a towing winch aft, which led to a record-setting towing job not long after she was built. The job was to tow a dry dock from the East Coast to Manila, through the Straits of Magellan, and it set a record for the longest tow in history up to that time. She was declared surplus after World War I, when the Griffiths Steamship Company bought her and put her in the Winslow shipyard for outfitting as a coastal trader. The ship was modified there to burn oil instead of coal, which meant adding fuel tanks, and longer cargo booms replaced the shorter ones used by the Navy to make the ship suitable for handling lumber.

THE *CAESAR*

The skipper of the *Caesar*, Captain John ("Jack") Clark, was a friend of my father's, a very nice fellow who used to come down to the *Bainbridge* often to chat with us deckhands aboard. While he was there one time I asked if there was any chance to ship out with him. He told me to report next morning to the mate. When I did, the mate said Captain Clark had already told him about me, and I was hired as an able-bodied seaman (AB), at $65 a month. Feeling pretty proud of myself I went home to tell Mama I was shipping out on the *Caesar*. She cried, of course, about her little boy leaving home at the age of sixteen years, but she accepted it and realized there wasn't really anything she could do to hold me back.

The ship carried six AB's, two winch drivers and a bosun for her deck crew. Although I did not have an AB certificate at the time, it was legal then for a ship to carry two uncertified AB's for every six sailors with AB certificates. At my age I was the "new kid" as far as the rest of the crew was concerned, and it didn't take long for them to put me in my place.

Going Offshore

On the first day I reported for work the crew had been washing down the decks, finding that one of the midship's deck scuppers was clogged so that the deck was awash. They soon had me over the side in a bosun's chair, carrying a wire to unclog the scupper. I was sitting right in front of the clogged pipe when it let go, giving me a bath from all the cold water that had been standing on deck. Most of the crew was watching when this happened, having a good laugh at my expense. Captain Clark was among them, and he added to the others' comments that it was a good thing I hadn't been cleaning out the outflow of the sanitary pumps.

We stood by the ship for about a week in Winslow before leaving port. The first thing in order was to swing the compass. This was before gyroscopic compasses were in general use, so the magnetic compasses had to be adjusted after a ship had been laid up for a time or had gone through extensive repairs. As soon as the *Caesar* cleared Eagle Harbor I was called up to the bridge to steer. With experience only on smaller boats with direct steering, this was the first time I had handled the wheel of a ship with power steering. It's possible to get the feeling of the rudder with hand gear on the small boats, but with power steering there's no such feeling and steering is done by watching a rudder indicator, which is quite often a bit off center. I did pretty well, though, and was in my glory as a new sailor going to sea for the first time!

Our first stop after adjusting the compass was at the oil dock at Point Wells. We had taken on just enough oil at Winslow to get that far, so now those new fuel tanks were filled for the first time. Next we went to Bellingham to load railroad ties, the first commercial use of that vessel. Bound for San Diego, the ties were loaded as high as fifteen feet on the deck, lashed down with chains turnbuckled down tight. Then we headed out the Strait of Juan de Fuca in nice weather, only gentle seas running.

It was an easy trip down the coast. As a member of the deck crew I took watches steering, and during daylight hours we scrubbed and painted various parts of the ship. She was powered by a triple-expansion steam engine that turned only 55 revolutions per minute, making for little vibration and comfortable running. As we got further down the coast the sun came out and the weather warmed, so I thought "Boy, I've sure got things going my way now."

The *Caesar* was a good feeder. Meal times and the kind of food served were pretty much standard on the ocean-going vessels of the time, but the quality and amounts varied a lot. On the *Caesar* the work day began with a breakfast of ham and eggs, hot cakes with syrup, cooked cereal, and a salted fish. Codfish tongues boiled with potatoes was a popular dish. Lunch was a hot meal with soup, meat such as roast beef or chicken, potatoes and vegetables. Dinner was similar to lunch, an even bigger meal that often included steaks or chops. A midnight lunch was served for the crew on watch, usually a cold meal while at sea and a hot meal while in port working cargo.

When we arrived in San Diego the ship was offloaded by longshoremen. Usually in the coastal trade the cargo was worked by the ship's deck crew, but in this case we just did our usual work around ship while watching the process. I was amazed to see how much cargo the ship had carried when the ties were loaded onto flatcars that more than filled the railroad yards at the docks.

The ties had been stored everywhere possible aboard, including some that were put under the poop deck. After these were removed we opened a door to some unused quarters that had been blocked by the ties. As the door was opened the ship's cat tore out of there and jumped over the side onto the dock, disappearing forever. The poor cat had been locked in there ever since the ship was loaded in Bellingham, meaning it had gone without food or water for about ten days. He must have thought that was really a starvation ship.

We worked only during the days while in port, giving us time to go up into San Diego in the evenings. On a Sunday, when we had the day off from work, we went up to Balboa Park and cruised around the city. I thought life was really wonderful! Riding the streetcars was a new experience for me. Unlike the streetcars I was used to in Seattle these were open, with seats along the side. I was also impressed with the huge number of Navy surplus ships still tied up at San Diego five years after World War I. There was a row of four-stacked destroyers moored alongside each other that looked like it stretched for a mile.

After discharging all our cargo we left for San Pedro to take on fuel, then went to San Francisco to load general cargo for the return trip north. That was a sad experience for me. I went ashore with some of the old timers who knew where there was a "blind pig," a place where you could buy alcohol during prohibition days. They bought some wine and egged me on to drink it. I wasn't very reluctant, as I thought that was part of being a sailor, so I got pretty drunk. The next morning I felt awful and didn't want to get out of my bunk, skipping breakfast. I was still in the bunk when it was time to turn to. The big German mate wasn't going to have any of that, though. He came down and yanked me out of my bunk, saying: "If you're going to be a sailor, kid, and go ashore with the gang, you'd better learn to get up in the morning afterwards. You get out on the deck. We've got work to do." I felt awful but I went out on deck and suffered through the day. I didn't go out and drink wine with the crew for a long, long time after that.

The return trip to Puget Sound wasn't quite as pleasant as going down the coast, with fog most of the way. The fog did give me my first chance to watch radio bearings being taken. The *Caesar* carried a radio operator who, under the directions of the mate on watch, would send a signal to a radio bearing station on the coast. The station would take a bearing on that signal, then take another two or three hours later. By triangulating those bearings the watch officer could determine the exact position of the ship.

Rounding Cape Flattery it was still very foggy, and there we used a time-honored method of finding the depth of the water. The ship carried a deep-sea lead and wire on a small reel aft, but these had been lost during the trip. So I was put to work sounding by hand, using a 20-pound lead which I had to take forward to the focsle head, passing it around all the rigging and stanchions. When I let it go I would sing out "lead away," then run all the way aft to help the mate on watch haul it in. The lead was armed with soap, so if it hit bottom we could see if sand, mud or some other kind of bottom was under the ship to compare it with the charts. I kept that up during all of my four-hour watch, which was pretty tiring.

Some of our general cargo was discharged in Seattle, most of it in Tacoma, then we went to Bellingham to load for the trip south. Built for the Navy, the *Caesar* had more quarters than necessary for a merchant ship crew. Our deck crew was housed in the petty officer quarters amidships, which were luxurious by merchant ship standards and even included white-tiled showers. To give an idea of what a luxury that was, on some of the ships I served on later the only way we could get a shower was with a bucket.

We were in Bellingham on a Sunday, giving me the day off. Getting ready to go up into town, I took a shower in all that luxury and then, since only men were aboard, threw a towel over my shoulder and walked naked back to my quarters. I didn't know that some Don Juan in the crew had brought a couple of girls aboard, students from the normal school in Bellingham. As I started down the alley here they came the other way. There was lots of screaming as they turned to run in the other direction. That wouldn't be such a big deal today, but in those days it was quite an incident!

I made two more trips on the *Caesar*. They haven't stuck in my memory the way that first trip has, the first of anything being more memorable than that which follows. I do remember that the big German mate didn't take to me very well no matter how I tried, so in early June I left the ship in Tacoma to go back to work for my father on *La Blanca*. My job on the *Caesar* wouldn't have been there much longer in any event. Soon after my departure the company changed her over to Canadian registry, putting an all-Canadian crew aboard. She was renamed the Mogul and stayed under the Canadian flag for a good many years.

We ran the usual berry operations with *La Blanca* all that summer of 1923. As it got to be fall the R.D. Bodle Company went into marketing huckleberries, many of them growing in the logged-off areas around the upper part of Puget Sound. We'd haul them into their cannery in Seattle, and also to the Olympic

Canning Company in Olympia. One day we'd travel down the Sound, picking up the berries at Gig Harbor, Warren, Arietta, Lake Bay, Home Colony, Longbranch and other small towns on Puget Sound where they had stores that gathered the berries from pickers in their area, and docks where we could land with *La Blanca* to pick them up for delivery to market.

Pickers went out in the logged-off country with sheets or pieces of canvas which they'd lay under the bushes, beating the bushes hard with sticks to knock the berries off. Of course a lot of leaves and other loose stuff would come off with the berries, so the stores had little machines with screens to separate the larger, ripe berries from the green berries, and blowers to blow away the leaves and other chaff. It's surprising how many huckleberries could be gathered with these small operations. By the time we got to Olympia after a day's run we'd have five or six tons of berries to deliver. We usually spent the night in Longbranch on the return run. The following day *La Blanca* would call at the same towns in reverse order, loading another five or six tons of berries for delivery to Seattle.

The huckleberry season ran until about the end of November. It was a nice, sociable run because of all the stops made at little towns along the way. When we lay in Longbranch, for instance, we'd go up to the store where people in the community gathered, telling stories of logging, fishing, and the sea. There was a popular game of the times played there, "Put and Take," played with an eight-sided spinning top. Players anteed a small amount, then spun the top in turn and followed directions on the top's upper side after it stopped spinning, such as "put five" or "take five." It was a gambling game of sorts, usually played with pennies so that no great amount of money changed hands.

Near Thanksgiving that year they had a big raffle in Longbranch with turkeys and geese as the prizes. I won two geese in the raffle, which Dad killed, cleaned and plucked for our Thanksgiving dinner. Mama being so religious, and opposed to gambling, we had told her about the spinning top game and raffle in Longbranch, but not that we played. The minister from her church came for Thanksgiving dinner that year. When he heard Dad and me talking about the game and raffle, he said it was a good thing the geese hadn't been won in a raffle; he wouldn't have been able to join in the meal in that case. I don't think he or Mama ever learned that their Thanksgiving dinner geese were the proceeds of gambling.

When the huckleberry season was finished not long after Thanksgiving, it was time for me to look for another job. Living at home while trying to find work, I spent a lot of time with a gang of friends. We'd go out to the dances at Fletcher's Bay and get together for parties. The parties always included some kind of alcohol, which of course had to be bought from bootleggers because prohibition was in force. Even if it hadn't been, our parties would have been illegal because most of us were very young. Though I'd already been to sea, I was only seventeen years old.

One night we were sitting around a cabin on the island, drinking beer. One of the fellows there, called "Kingfisher," had a glass eye that had a habit of popping out unexpectedly. In the middle of the party it did just that. We all got down on our hands and knees looking for it. The eye couldn't be found anyplace, so we went back to drinking our beer. Kingfisher was just finishing his beer when he felt something in his mouth. Spitting it out in his hand, there was his glass eye. As far gone as we were by that time, we all thought that was hilarious.

My mother didn't think it was very funny the next morning when she found me with an awful hangover. "You've been out drinking, haven't you?" she accused me. Though I denied it, she didn't believe me.

A couple of days later I was home when the county sheriff turned up at the door along with a deputy. Mama showed them into the living room and asked me to join them. "I understand you were out drinking, young fellow," the sheriff said. "Whatever you were drinking, you know you can go blind drinking that stuff."

"Who said I was drinking?" I didn't feel very comfortable.

"Well, I got the word that you were out there drinking. Now, I want you to tell me where you got that liquor. Who sold it to you?"

I told the sheriff I didn't know what he was talking about. I hadn't been out drinking. But he pressed his case hard. "You think you're a tough guy. We take guys like you, put them in jail and throw away the key until they confess where they got the stuff!" He stayed there for a couple of hours, threatening all the time while I kept denying I had been drinking. He finally went away in disgust.

Mama must have been awfully relieved when I got a job. My older cousin, Chris Christensen, had a Puget Sound Mate and Master's license, and a job with the Chesley Towboat Company of Seattle. They had several tugs, all steam-powered, including the *C.C. Cherry*, *Columbia*, *Tempest* and *Katy*. In early December of 1923 Chris, who was serving as mate on the 85-foot *Katy*, called to ask if I would like a job as deckhand on that tug. I took the job at $60 a month.

She was an interesting old boat. They had just finished cleaning the boilers and tubes. Even though she had condensers to conserve fresh water, those old steam tugs took lots of it. After we had filled with water and fueled up, we headed up towards Semelt Bay near Deception Pass to get a string of logs. With all that water and fuel aboard, the *Katy's* guard rails were just about even with Puget Sound, no more than two or three inches of freeboard remaining. She did gradually get higher in the water as the fuel and water were used out of the tanks, but she was a very deep vessel drawing about fourteen feet of water.

The crew was typical of towboat crews at that time, with a captain, mate, cook, deckhand and two engineers. Watches when under way were "six and six", meaning that a crewman stood watch from noon to 6:00 PM and from midnight to 6:00 AM, or the alternative six-hour periods. As deckhand I drew a wheel watch during those 12 hours a day with the captain, an old German named Charlie Naegler.

On the first of those watches, heading north in the Sound, I stood at the wheel for a straight four hours without a break, with Captain Naegler wandering in and out of the wheelhouse and sometimes standing in a corner for awhile, watching me. After those four hours I was getting tired, so I reached over and dragged a high stool in front of the wheel to sit down for a time. The captain was just waiting for that.

"These damn kids nowadays," he said, "they can't stand anything anymore, they just get tired and got to sit down."

"Well, Captain," I said, "if I could just go get a cup of coffee or something I could probably stand up for the next two hours."

"Go get a cup of coffee, and bring me one," he snapped, taking the wheel himself. That was my first run-in with the skipper, but after that we got along pretty well as he began to see that I was a good deckhand and had more experience than my age might indicate.

We arrived at Semelt Bay and got ready to tow those logs. They were first-growth, ranging from three to five feet in diameter, and we were to tow somewhere between 30 and 40 sections. Heading with that heavy tow down to Tacoma was a slow process. Those old-timers like Captain Naegler knew the Sound well, working the tides and taking advantage of eddies close to shore, but even with his experience it would take three or four days to get to Tacoma. On later trips it sometimes took even longer if bad weather came along and we had to lay up somewhere along the way.

I got so I liked that old boat and the slow pace of towing logs. The *Katy* was comfortable for that winter work, as all the old steam tugs were, because with the availability of steam there were radiators throughout the boat making it nice and warm. She was an almost historic boat even then, built in San Francisco in 1868 and brought to Puget Sound at some later date. Heavy oak planking made her a sturdy boat. She had been coal-fired until brought to the Sound, when she was converted to an oil-burner and her house was rebuilt along the lines of most Puget Sound tugs.

One of my jobs as deckhand was to take care of the kerosene lanterns carried on the log rafts, one to each corner. Usually that meant cleaning and re-fueling four lanterns a day, though sometimes it required tending eight lanterns if we were towing two rafts. I did this on my daytime watch. When the lanterns were ready I would put them in the tug's skiff, put the skiff in the water, drift back to the raft and exchange the newly fueled lanterns with those that had been hanging there for 24 hours. That required tying up to the raft and crossing it on the floating logs, a pretty dangerous procedure because one slip and a drop between the logs would have been the end of me. Once the exchange of lanterns had been made, I would row back from the raft to the tug, which never slowed during this time. Even though the tug wasn't very fast while towing log rafts, the row back from the rafts could be very strenuous, especially if a stiff wind was blowing.

Just as with *La Blanca* on the berry runs, there were some nice sociable touches to this kind of boat life. One of these came whenever we passed Eagle Harbor, crossing the run of the *Bainbridge* between Winslow and Seattle. Captain Welfare, still skipper on the boat, would come alongside us as close as possible, then toss a copy of the latest Seattle newspaper onto our deck. This was a welcome diversion and something Captain Welfare did for most of the tugs crossing his route. The tugs didn't carry radios in those days, so that was the only news we would normally get during the long tows down from Semelt Bay.

I hadn't been aboard very long before Captain Naegler decided to test my seamanship. He threw some rope on the deck and told me to put an eye splice into one end. I threw a nice, neat eye splice into that rope in no time, which amazed the old man. When he asked how I had learned that I told him about sailing with Dad, and Dad's long experience at sea and how he had taught me. Then he learned from my cousin Chris that I already had some deep sea trips behind me, and from then on we got along very well. That was only for a few weeks, though, as Captain Naegler got sick and had to be replaced.

Captain Bill Stark was a much younger man than Captain Naegler, probably about forty-five years old when he came aboard the *Katy*, a much different kind of skipper. He had been a rum-runner, smuggling whiskey from Canada down into Puget Sound using high-speed power boats. He had many stories to tell about those days, being especially proud of some wonderful binoculars he used to spot the revenue agents. They were such good glasses, he claimed, that with them he could see the patrol boats bright and clear at night. Along with his whiskey smuggling experience he had a thirst for the stuff. He didn't limit his drinking to times ashore. While he was drinking on the tug he often didn't feel like standing a full watch, so when we were on watch together he quite often would trust me to navigate while he slept.

His quarters were just aft of the pilot house. On these occasions he would shout in to ask me "Where are we now, Holger?" After I had told him he would say: "Okay, if you're going to run aground let me know so I can get my clothes packed." Then he'd roll over and go to sleep again.

On one trip we were towing a raft into Lake Washington, which required breaking it into two parts so it would fit through the Ballard locks. We anchored out in Shilshole Bay off the locks, preparing to divide up the raft, when Captain Stark decided he needed a drink. "Come on, Holger," he said, "let's take the ship's boat ashore and go up into Ballard." We did, and he left me standing on a street corner while he found a bootlegger and got himself a bottle of booze. He came back happy as could be, and we went back to the tug to split up that log raft.

That was quite a job. It was a very cold January day. We had a new mate aboard, my cousin Chris having left to skipper another tug. The new mate was a huge fellow named Rasmussen, a capable officer but he had lost his equilibrium as a result of being hit on the head with a booze bottle. Splitting up a boom like that is hard work, involving slipping the chains on one section, tying it up to one of the moorings in the bay and getting under way with the tug towing the rest of the raft. "Rass," as they called the new mate, proved to be no help at all, as he would get on a log and freeze there, unable to move.

Somehow we got it all together, and as we were moving through the locks and up the canal the old timers were reminiscing about some of the dangers of working log rafts. One of their stories was about Captain Naegler. While tied up with a log raft on one of his trips, he had a row with his mate at that time, a big fellow who got so mad that he took a peevee and chased the captain right down to the end of the log raft. Old Charlie got to the end of the boom, turned and shook his fist, saying, "I go no further!" He didn't have much choice; he'd have been in that cold water if he'd gone any further. With no radio and little to read, such story-telling was one of a crew's main diversions.

The Chesley Towboat Company turned out to be a disappointment for me when they failed to come up with my salary. After they had missed two pay periods I threatened them with a lawyer, to which they countered that they would have me blackballed. I didn't have much choice but to quit at that point, going back home and making a couple of trips with my father on *La Blanca*. But he didn't have much work for me since it was still winter, and he had a regular crewman, a Finn named Harry Bowman.

I was at loose ends until the shipyard called and offered me a job with their boilermakers to fit new tanks into the *Fern*, a lighthouse tender that belonged to the U.S. Lighthouse Service. My first job was to help cut out plate holes, then buck red-hot rivets to assemble the tanks when that work was finished. Next, I helped the pipefitters install the tanks in the hold. While working around there I asked the mate of the *Fern* if I might have a chance to ship out with them, telling him of my experience up to that time. One of the crewmen quit shortly before the job was finished, so I was given a job as an AB. I was pretty proud until discovering that because I was only seventeen years old I had to get my parents to sign a permission for me to ship out on a federal vessel. That took me down a peg or two!

We left Winslow in April of 1924, going first to Bell Street Wharf to take on supplies and then heading up the Inside Passage to Alaska. That was quite a thrill, my first trip to Alaska since leaving there as a small child. The *Fern* tended the lighthouses in Southeast Alaska, based in Ketchikan. We set buoys, recharged lights and took care of other navigational aids as far as Cape Spencer, similar duties westward in Alaska handled by a larger ship, the *Cedar*.

The *Fern* was a nice little ship, steam-powered, comfortable and warm. We kept busy all the time and I enjoyed the work, although it was hard at times. The unwatched lights were lit by an accumulator, charged by four gas cylinders. Each cylinder weighed about 120 pounds. When replacing them we'd have to throw one of those cylinders on our shoulders and carry it up to the light, quite often located up a steep hill on a small island. The first few times I did that I was pretty winded, but after awhile I got used to it and could do it along with the rest of the crew. I was always amazed at one of the crewmen, known as "Frenchy," a little fellow who couldn't have weighed as much as one of those gas cylinders. He could handle those cylinders as well as the biggest ox in the crew.

Some of the lighthouses were unmanned, working automatically, while those that were manned were classed as "married" and "unmarried" stations. The *Fern* was sometimes called out on medical emergencies to those manned stations. One such run I remember was to Tree Point, south of Ketchikan, where the wife of the light keeper was pregnant and soon to give birth. It was very rough weather as we ran down there, going ashore in the ship's lifeboat. By the time we got to the lighthouse the woman was in labor. We got her into the lifeboat and back to the ship, setting sail at once for Ketchikan. It seemed we would never get there before the baby came, but it held off long enough to be born in the hospital in Ketchikan.

Miscellaneous jobs like that made life aboard the *Fern* interesting. Another example was when a high school track meet was to be held in Juneau. The team from Ketchikan had no way of getting up to Juneau, so special permission was granted by the Government to transport them on the *Fern*. Since it was a 24-hour run each way to Juneau and back, we had a big gang of high school kids sleeping all over the ship.

In July of that year one of my shipmates, Charlie Anderson, left the *Fern* for a job on a cannery tender running out of Ketchikan. After one run he came down to the *Fern* and talked me into sailing with him on the tender. It was a good job, he said, and at $90 a month the pay was a lot better than the $75 we were earning on the *Fern*. I spent the rest of that fishing season on the *Leonine* along with Charlie, building fish traps, brailing them and hauling salmon to the cannery. The skipper of the *Leonine* was Larry Parks, a fellow I saw often in later years. He eventually got a license and worked for the Alaska Transportation Company as captain of the old *Tongas* on the Seattle-Alaska run. When World War II came along he was commissioned in the Naval Reserve, serving as captain of an oil tanker. We ran into each other often in various ports and became good friends.

Those were fine days on the *Leonine*. She belonged to the Sunny Point Packing Company in Ketchikan. The canneries were noted for feeding well and the boat was comfortable for the crew. While laying in port at Ketchikan we had a chance to get ashore, going to occasional shows and visiting friends around town. I really enjoyed the days spent there, even if it rained all the time. Nobody seemed to mind the rain, though, and I remember it as a good time in a great part of the country.

At the end of the season in late September I returned home to Bainbridge Island, again working some jobs with Dad on *La Blanca* while looking around for another seagoing berth. I couldn't have known when I found that berth in November 1924 that it would prove a transition in my life, with many lessons both good and bad for a future at sea.

CHAPTER SIX
Griffson

THE *S.A. PERKINS*

IT WAS IN THE LATTER PART OF NOVEMBER 1924 that I went down to the *Griffson* in Eagle Harbor and got a job as an AB. Soon after I signed on, she was towed to Port Angeles to load lumber at the Charles Nelson mill, a cargo bound for Oakland, California. After some days of loading we were taken under tow by the *S.A. Perkins*, a big steamer also owned by the Griffiths Company, fitted with two towing winches aft for the purpose of towing a number of barges owned by the company.

It was nasty weather when we headed out the Strait of Juan de Fuca, with a stiff southwest wind blowing. By the time we got to the open ocean beyond

Cape Flattery, the wind was blowing so hard and the sea so rough that the *S.A. Perkins* turned around and came back into the Straits, quite a feat of seamanship considering the height of the seas and our barge in tow. We were left at anchor at Clallam Bay while the *S.A. Perkins* headed back out to sea and down the coast.

By now it was almost Christmas, and pretty obvious that our holiday would be spent at anchor there awaiting another ship to tow us to Oakland. Captain Clark sent a couple of us ashore in a small boat to buy some chickens for our Christmas Day dinner. We managed to find the chickens and also were able to locate the local bootlegger to buy some moonshine. Those chickens were old and tough, but the cook stewed them tender for a wonderful Christmas dinner. We had a happy time even if we were stranded there at anchor.

After Christmas, we did odd jobs around the barge. I was lucky to have been appointed "donkey man," charged with running the big donkey boiler located forward on the barge. The donkey was fired by a heavy crude oil, too heavy to use while it was cold. So I would have to start the fire with wood. After the wood fire had got up steam in the boiler, the steam was run through coils in the oil tanks, warming and thinning the oil so that it could then be used for firing the boiler. I liked that job, which also included maintaining the winches and other gear associated with handling cargo. The ship had two sets of cargo winches to serve her four hatches. Two winches were in each set, one for the yard arm and one for the boom. Each winch was self-powered by steam fed from the donkey, and the steam was also fed to the anchor windlass as well as through a long line running aft to a large capstan for handling lines.

We waited there until just before New Year's Day, when a message came that the *Griffson* would be towed by the steam schooner *Mukilteo*, a lumber carrier belonging to the same Charles Nelson Company at which we had loaded. When we saw her coming I fired up the donkey to get up steam for raising the anchor. Preparing a large tow like that for sea was quite an undertaking. The main towline from the *Mukilteo* was a two-inch cable, which we made fast to anchor chain on the anchor windlass. We'd then pay out about 40 fathom of this chain. The weight of the chain acted as a damper to prevent surging in the ocean swells. Towing directly on the anchor windlass like that would have put too much strain on its fastenings so we also used a "devils claw," a large metal hook that engaged with the anchor chain where it came off the windlass and was held by a heavy cable run around the foremast. All these precautions helped prevent snapping the tow line or tearing the anchor windlass out of the deck, and were quite typical of setting up a deep-ocean tow in those days.

Just before the *Mukilteo* went ahead to take up strain on the tow line, the Finnish captain yelled across at us: "Don't be afraid if you hear whistles blowing tonight, it won't be a distress signal. It's New Year's Eve that we'll be celebrating." That memory makes it easy for me to place the date of heading out to sea for the first time on the *Griffson* as the last day of 1924.

It was an uneventful four or five days towing down the coast to San Francisco Bay. Just off San Francisco the *Mukilteo* dropped the tow line. We hauled in the anchor chain and line, then a towboat made fast alongside to take us in to the Charles Nelson lumberyard at Oakland. After discharging the lumber we were next towed to the Southern Pacific docks, to take on a load of redwood railway ties for Guaymas, Mexico. We lay there for quite some time, as the loading was done from small coastal steamers coming down from Eureka and Arcadia loaded with the hand-split ties. After we had a full load from several of those steamers, we had to wait for a tow to Guaymas. The ship that finally came to take us in tow was the *S.A. Perkins*, the same steamer that had dropped us in Clallam Bay at the start of the voyage. She towed us out through the Golden Gate, down the coast, around Cape San Lucas, and back up into the gulf to Guaymas, on the mainland of Mexico. There we dropped anchor until some small tugs towed us in to the docks at Guaymas.

Everything seemed to take a long time on such a trip. We had to wait quite awhile before the tugs took us in to the port, and then the unloading was terribly slow. Working only during the daytime, it took more than two weeks to load the ties onto railway flatcars. Every day we had to take a water barge alongside to replenish our supply. The donkey boiler had no condensers, discharging the steam directly over the side and using a lot of water.

But it certainly wasn't a case of "all work and no play." We met some families in Guaymas and had some wonderful times with them, going on picnics on Sundays, meeting lots of pretty Mexican girls. With such a small crew we were able to have them come down to the ship for visits, which were fun but very chaste. The girls all had to have chaperones along with them in those days. After the decks had been cleared by unloading we even held a dance on the *Griffson*, hiring an orchestra of four or five pieces for some 15 pesos, dancing right on the deck.

On one Sunday the families invited us out to a farm in the countryside. Somebody had an old Model "T" Ford truck, loading all of us into the back and driving twenty or thirty miles inland to the farm. There they had kegs of Mexican beer which was drunk in great quantities, the weather being so hot and all, and had arranged a turtle feast for us. They had big sea turtles which were propped up in their shells around a bonfire, leaving them there until the shells came off, indicating that the meat was done. Then the meat was mixed up with a lot of spices, making a wonderful dish that was very hot and needed more beer to wash it down. Everyone was feeling pretty good by the time we piled into that

Model "T" for the return trip. We really had a grand time with those fine people.

There were three or four of us the families took a particular liking to, inviting us to their homes in the evening. There would be a piano in the home, providing music for dancing, and we'd have a little party each time we went. I took a special interest in one pretty little Mexican girl, but could never get anywhere with her because of her chaperone. We'd try to ditch the chaperone time after time, with no luck at all. It was hard getting used to those customs.

There was a park with a bandstand in the center, where they had music each evening and people went walking. We'd go there sometimes, but found it useless as a place to get together with the girls. The men paraded one way around the bandstand, while the women counter-walked the other way. We'd walk along with the men in their direction, keeping an eye out for the girls, but then when we stopped to talk with them we couldn't do more because they'd have to continue in their direction and we in ours.

The exchange rate at that time was two pesos per dollar. We were allowed to make ten-dollar draws on our pay, which were exchanged for twenty silver pesos. It made for a heavy load in your pockets, but it bought an awful lot of stuff for the amount it represented.

One of the uses for that money was in the beer parlors in town. We drank mostly beer at that time, and spent many of our free hours in those places when we weren't with our family friends. Captain Clark did more than his share of that kind of drinking, linking up the first day we were in port with an old German fellow called Doc Frieban, a man who was really just a beachcomber. About the third day in port I was working in the donkey boiler room. The combination of weather and heat from the boiler had made it so hot that I rigged up a wind chute above one of the skylights to catch the breeze, hanging some bottles of beer in the chute to cool them. Suddenly Doc Frieban stuck his head in there, rubbing it and saying "I got a head on me like a bushel basket today. I was out with your skipper last night and we overdid it." Since I hadn't seen the skipper for a day or two I could believe he was feeling the same way.

We had a Swede aboard ship who was a bit of a loner, known for dressing up in fine clothes when he went ashore. On one trip ashore in his fancy clothes Ole got himself pretty full of drink, into some kind of trouble, and wound up in jail. The next day they called Captain Clark to come and bail him out. When the captain got there he paid the fine the police clerk requested, but then the clerk said "that's not all." "What else?" asked the skipper. "Well," said the clerk, "he's an officer on your ship and we couldn't put him in with the rest of the guys. So you have to pay for a blanket, and a nightshirt, and these other things." He gave Captain Clark a list of all the things Ole had received as part of this special treatment, which the captain had no choice but to pay.

Like all good things, our stay in Guaymas came to an end. After we had been some twenty-five days in port the *S.A. Perkins* returned to take us in tow again, running across the Gulf of California to San Marcos Island to load gypsum. The Griffiths Company had a share in an operation run by the Standard Gypsum Company, mining gypsum from this island composed entirely of the stuff. Both the S. A. Perkins and the *Griffson* took on full loads of gypsum, put aboard by conveyor belts. Loading was a dirty job, the dust getting everywhere aboard ship. Afterwards we'd have to clean out all the living quarters, even our clothes lockers where the dust would seep around the doors.

The *S.A. Perkins* then towed us to Long Beach for unloading, while she continued on up to Puget Sound to discharge her load at the Standard Gypsum Company's facility at Harbor Island in Seattle, where the Todd Shipyard is now located. After unloading the gypsum at Long Beach, a tug towed us to Eureka where we took on another load of redwood ties. Meanwhile, the *S.A. Perkins* took on general cargo around Puget Sound and then picked us up at Eureka on the way down the coast. She again towed us to Guaymas, then went to discharge her cargo at ports in Central America and the northern part of South America. It was a well-coordinated operation, scheduled so she would get back from there to Guaymas after the twenty-four or twenty-five days it took us to unload the ties, then both ships would be loaded again with gypsum for Long Beach and Puget Sound.

We continued on this run for many months with a few variations to the schedule. Sometimes we were towed to Oakland rather than Eureka after offloading the gypsum at Long Beach. One of the more interesting changes began at Guaymas during our second trip there, when the local tug was late arriving to tow us over to San Marcos Island for the usual load of gypsum. Captain Clark, tired of waiting for the tug, ordered us to break out the sails. We pushed clear of the dock at Guaymas, hoisted sails and sailed all the way over to San Marcos Island. We moved at only about two knots, and had a tug assist us in getting to the pier at the island, but it was a thrill to be under our own power for a change.

On that same trip, the *S.A. Perkins* was unavailable to tow us north after loading gypsum. Instead, one of the big Red Stack tugs from San Francisco came down to put us in tow. The *Sea Scout* was one of a fleet of big, powerful steam tugs carrying the Red Stack colors, well known up and down the coast. I was amazed to see that she hooked us along at seven or eight knots on that trip north, quite a speed for the size of our barge and the full load of gypsum she carried.

This was a fine job for a number of reasons. One was the special position I had on the *Griffson* as donkey man, placing me a shade above the other seamen.

I had my own room on the ship rather than bunking with the others in the focsle, and I was in charge of the work when the captain was ashore, making me feel like a big shot. Another was the friendships we had made in Guaymas, renewed each time we returned. There were more picnics, more parties in the homes, and more times spent at the park.

But the most important reason was my growing friendship with Captain Clark, who became my mentor in work and play, both good and, I have to admit, bad.

Captain Clark was an excellent seaman who at that time was still struggling with a drinking problem. His appointing me as donkey man on the ship was an example of his interest in my career, also shown in his leaving me in charge when he was away from the ship. He was responsible for my getting an AB certificate, prompting me to take the examination during one of our layovers in Oakland. He gave me a letter outlining my time and duties on the *Griffson*, which I took over to the United States Steamship Inspection Office in San Francisco, along with some other letters I already had from other ships. I made out the application and then went into the examination room, where there were about a dozen young men taking examinations for various tickets.

An old captain giving the examinations started asking me some questions, then noticed on my application that I was born in Nome. He lit up at that, since he'd spent some of his younger years there, and started talking about his adventures. One of the men taking an examination raised his hand to ask a question. The captain told him in a gruff voice: "Can't you see I'm busy up here? You just wait until I get through." The hand went down while he continued to tell me about his experiences in Nome. When he was finished he gave me my AB ticket. I left there pretty proud of myself, knowing that I was a full-fledged seaman. At that time the AB certificate was the only one given to seamen, entitling the holder to sail in any unlicensed position including bosun and quartermaster.

As a mentor for play, Captain Clark's influence wasn't of the kind I would have told my mother about. His thirst for drink went beyond the ship's budget so far that he was always involving me in some scheme to raise money for our "social activities." Once, in Oakland, we went through the railway yard collecting all the stakes from the flatcars and loading them on the *Griffson* wherever we could find room. When we got to Guaymas he sold them and split up the money with the crew for our fun ashore. A couple of other times, late in our stay there, he got me to help him take paint and other things from the ship's stores to sell up in the town for more booze money.

With prohibition in the States, the skipper also wanted to make sure of a drinking supply when we returned there. His solution to that was to buy ten cases or so of tequila in Guaymas, which we would hide way back under the stern quarter in some big beams. He'd do that when he knew we were going back to Oakland, as a port captain in San Francisco was a good friend with whom he wanted to share the loot. When we were tied up at the Southern Pacific dock in Oakland, he would load up a suitcase full of tequila and take me with him to carry the bag, figuring it would be better for me to be caught than for a ship's captain. We'd walk across through the depot to catch the ferry to San Francisco, not buying tickets because we were already past the ticket control where we were docked. On the other side we'd walk to the Pacific Steamship dock to visit Captain Clark's port captain friend. The company's port captain was a former skipper of coastwise ships for the Pacific Steamship Line, son of a captain on the old President Line ships, and had become good friends with Captain Clark while both were sailing the coast. He and a number of friends would be sitting around the office, where Captain Clark would take the bag from me and pass the tequila around.

We also used to visit the local bordello in Guaymas, known as "Madame Teche's." It was one of those genteel houses, genuinely a place to visit, drink, and socialize with the local people without feeling forced to go to the back rooms with the girls. Madame Teche's was located in the downtown part of Guaymas. From the outside it looked much like all the other one-story adobe buildings. The front part of the building was a beer parlor, nicely appointed and the best in town. But the real difference between Madame Teche's and the other places was in the girls who hung around the beer parlor. They were young, very pretty, and available. None of them I met spoke much English, but that wasn't a requirement in their business.

I have never been able to understand one of the oddities of doing business with those girls. As all over the world, they demanded their payment in advance, in this case it was made in silver pesos after going back to the room. Each and every one of them then proceeded to wash the money before turning the trick. I asked the girls I knew about the reason for that, as did all my friends who visited the establishment, but they refused to give an answer. Since they were good Catholic girls despite their profession, the only explanation I have for this strange procedure is that they somehow considered their payment as "dirty money," going through a ritual washing of it before getting down to business.

During my later visits to Guaymas I had gotten close to a Mexican lady who was married to a politician. Her husband spent a lot of time in Mexico City and we had fallen for each other in his absence, truly an affair of the heart. I still blush to remember being at Madame Teche's one night when a young boy arrived with a message from my lady friend, who had tracked me down there. "There's nothing you can get there," she had written in English on a note handed to me by the boy, "that you can't get here!" We corresponded for some years after I left that job on the *Griffson*, but the letters eventually tapered off and we lost contact.

Captain Clark's friend, Doc Frieban, turned out to be an interesting fellow even if he was a beach bum. He had been a civil engineer and had been involved in a number of mining ventures with a partner who still had an office in Guaymas. Old Doc was quite a character, with many interesting stories to tell of his adventures. During one of our times in Guaymas, he took me to his former partner's office because there was something he wanted to show me. He told me they had been millionaires at one time, and to prove it took me to a closet that was completely full of Mexican pesos in paper money. The story was that the two partners had great success with one of their gold mines, but just when it was going good the mine was nationalized by the Mexican Government. Before they could bank all the money from the sale of gold and what they got from the government for giving up the mine, that particular issue of pesos was declared worthless. The old gentleman who was his partner verified the story, so I believe it was true.

Despite his comedown in life, Doc Frieban was still quite capable in many ways. During one trip to Guaymas, our donkey boiler was in bad shape with the tubes going out. Captain Clark hired Doc to help me re-tube the boiler, a hard job involving cutting out the old tubes with a cold chisel, putting in the new tubes, reaming and re-beading them. It took a long time to do the job. Doc Frieban was a great help, and it helped him to earn a few pesos.

After many trips like these on the *Griffson*, sometime in June, I found a letter awaiting me in Oakland. My father wrote that he was getting busy again with the berry business and wondered if I might be able to help him out through the summer. I hated to leave, and Captain Clark didn't want to see me go, but he understood that I had a deep feeling for my father and felt I should return to crew with him on *La Blanca*. To get back to Puget Sound I took a "workaway" passage up on the *Emma Alexander*, supposedly working in return for not paying fare, but actually not having to do much. My new AB ticket helped in getting that passage. I was feeling pretty good about things when I got home.

CHAPTER SEVEN

More Years on Deck

THE *ODUNA*

MUCH AS I ENJOYED my time on the *Griffson*, it was good to get home. This summer of 1925 was a busy one for Dad's operations on the Sound, involving almost all our family.

With all the berries to haul, the blasting materials to carry for Du Pont, and other jobs he had contracted, Dad had me running *La Blanca* that summer with my next-younger brother acting as deckhand. Paul was then fifteen years old, a high school student already showing signs of growing into a big, tough adult. He and I worked *La Blanca* on the explosives runs for Du Pont, on the odd towing jobs, and helped out on the berry runs when not off someplace else in Puget Sound. Although I didn't need a license to operate the boat for these jobs, I did need one if we ever were to carry passengers. That summer I got my operator's license in Seattle, permitting me to skipper any boat sixty-five feet or under in length and the first of a series of licenses I was to get in the coming years.

Dad, meanwhile, had chartered the purse seiner *Saint Michael* to work the berry runs. The boat's owner was Salvador Criscolo, who went along with the boat as engineer, while my thirteen-year-old brother Nels was deckhand.

My father's business was truly a family operation. The only male member of the family not involved was my oldest brother, George. At the age of twenty-one, George was already married and had two children. He had also been to sea, working for the Griffiths Steamship Company as an oiler on the steamship *Grifdu*, and was now serving as engineer on a yacht.

With Dad, myself and my brothers on the boats, Mama and my sisters took care of feeding us. Hannah was a high school student of seventeen, while Ellen was nine years old and still in grade school.

A typical day for the berry run began in Seattle about 5:00 AM, when we would get up and have breakfast on the boat. Our first stop was to load sugar at Pier "D," the Old McCormack dock as it was known then. Next we'd make a run for empty barrels, usually loading them from boxcars at Pier 4, sometimes going in through the Ballard locks to the Western Cooperage Company on Lake Union when a full load of barrels was needed. We arrived at Winslow shortly after lunch was eaten aboard the boat, then took a nap until about 5:00 PM, when all the crew would row into shore to have dinner at home. Besides the men in the family there were anywhere from three to half a dozen additional crew members hired by Dad, often including my old shipmates Jim Murray, Smokey Hansen and Bill Henshaw. Mama and the girls had a big gang to feed. I sometimes felt sorry for them because it was so hot in the kitchen with the old wood stove blasting away, and such a hungry lot of men to feed. But Mama seemed to enjoy being part of the activity, and my sisters gave her a lot of help.

After dinner the work of loading fresh and barreled berries began. By the time they were loaded, hauled to Seattle and discharged at Pier 4 and the cold storage at Spokane Street, it would be midnight or later. After a midnight lunch of cold cuts, cheese, bread, and coffee cake, the crew would fall into bed for a few hours sleep before starting over again.

Once again the fall of the year saw all those activities tapering off. It was time to look for other work. I was just starting to think about where to begin when a call came from Jim Murray, who had gone to work on the *S.A. Perkins* along with another friend from Bainbridge Island, Harry Martin. It was natural for them to think of me when one of the crew quit shortly before the ship was to sail. "If you get over here to Seattle before two o'clock this afternoon," Jim told me, "you've got a job as an AB."

I didn't waste any time throwing my things together and getting on the Winslow-Seattle boat. The ship was fully loaded and ready to sail for Panama when I arrived, the crew putting final lashings on the deck load. When I went aboard ship the mate, Mr. Bunker, took a look at me and said "My God, you got any more like that on Bainbridge Island? I'd do all right if I could get a crew of you fellows!" I was a big, strapping young man, over six-feet two-inches tall and weighing about 200 pounds. Harry Martin was about the same size, and though Jim Murray wasn't quite as tall he was very stocky and strong. Mr. Bunker must have thought that they grew a special breed of young men on the island.

I enjoyed the trip south immensely. She was a nice ship, the *S.A.Perkins*, a large, modern steamer with five hatches. Mr. Bunker was a pleasant mate to deal with, and the skipper was equally pleasant. He was Captain Earl Mercer, formerly a mate on the same ship and now sailing for the first time as skipper. The icing on the cake was having my two old friends from Bainbridge Island aboard.

This was the first ship I'd been on that had an "iron mike," the name given to an automatic steering device run off a gyroscopic compass. That cut down the crew needs on the bridge, since steering watches could be handled by less experienced ordinary seamen with the AB's steering only when entering port or in other situations where good steering was required. Because we were relieved of steering watches, the AB's worked only during the day on an eight-hour shift. When warm weather was reached in more southerly waters we started painting all the topsides. As the youngest member of the crew my first job was to paint the mast. This involved being hauled to the top of the mast in a bosun's chair, fastened to a gantlin line run through a block at the top of the mast. My paint and brushes were on a separate, smaller line. Once I was hauled up I could lower myself a little at a time by taking a loop with the line, hanging on with one hand and throwing the loop over the top of the bosun's chair and under my feet. With the sun shining, the tropical weather and a gentle sea running, that was a nice job.

During the two weeks it took us to run to the Panama Canal we were able to do so much painting that our ship looked like a yacht by the time we arrived. Discharging the cargo was a very slow process, taking about three weeks, so we had time then to finish up painting by freshening up the sides and taking care of those deck areas we hadn't covered on the way down because of the deck load.

We had time, too, to visit ashore, my first experience in a foreign country other than Mexico. Fortunately, my friends and I didn't get into trouble, which cannot be said for all the sailors in that port. One of the Panama Canal railway ships, the *Panama*, had been purchased by the Alaska Steamship Company for use on the Seattle-Alaska run, and was in dry dock getting ready for the trip north while we were there. She was a large, coal-burning ship which required a lot of firemen. There were about twenty or more brought in to man the boilers on that ship, a tough bunch of guys known as "Liverpool Irish." On their very first night in Panama they went up town and got into a fight, with the whole bunch of them winding up in jail.

The skipper of that ship was an old Alaska Steamship captain named Glasscock. The next day the Panama authorities called the captain, asking him to come bail out his crew of firemen. "No," said Captain Glasscock, "you keep them there until we sail." He was told that the authorities couldn't do that; with such a large crew it was costing too much to feed them. "We'll pay the board bill if you keep them there," said the captain. And that's just what happened, the crew staying in jail right up to the day the ship took off, headed north.

My friends and I spent most of our time sightseeing, exploring the streets of Panama City and seeing an occasional show. The sight I remember best was the ruins of a beautiful old church built by the Spanish in 1537 and said to be the oldest in the Americas. It had been partly destroyed by the pirate Henry Morgan in 1671, along with most of Panama City. What I remember best is the story about its original construction. Twice during the building of the church the roof caved in. As the roof beams were being placed for the third time, the priest knelt under them and prayed that they fall on him or remain there forever. The roof stayed in place and the priest lived, though neither one of them forever.

Leaving Panama we went north to San Marcos Island, where I had been so often before on the *Griffson*, to load gypsum for Long Beach. After discharging at Long Beach it was a big job to clean the holds for general cargo, then we proceeded to Portland. It was January 1926 when we arrived at the Columbia River, the coldest January I remember in that area. The river was entirely frozen over with three or four inches of ice. We plowed through that ice all the way from Astoria to Portland, where we loaded grain and sack goods for Central and South America. The loading had to be planned carefully so that the cargoes bound for different ports could be offloaded in order. Much of what we loaded in Portland was placed and shored up in the 'tweendecks, a storage area above the main holds. The remainder of a load of general cargo was to be taken on at Seattle.

We left Portland during the worst storm of the year. Reaching the Columbia Bar, the seas were so large that several ships had laid up at anchor waiting for the storm to abate. But Captain Mercer was an old coastwise sailor with pilot's papers for all the river entrances; he wasn't about to let the storm hold him back. As we crossed the bar the seas were breaking so heavily that green water was coming over the bow, all the way back to the midships house. I saw the engineer shaking his head. "If anything happens down there," he said, nodding down at the engine room, "its curtains for us." I imagine that half way across the bar Captain Mercer wished he hadn't tried it, but we got across safely and started laboring up the coast, the ship rolling heavily all the way up to Cape Flattery. When we got to Seattle and opened the hatches, all the cargo that had been shored up in the 'tweendecks had fallen into the lower hold, the shorings broken by the beating the ship had suffered in that weather.

It was a mess, but we eventually got it cleaned up, took on the remainder of our load—some piling and lumber—and were ready to sail again. This time we were headed for La Union, in San Salvador; then on to Panama, Ecuador and Peru. After taking on fuel at San Pedro on the way south, our first call was at La Union. While some general cargo and lumber was being discharged we had a chance to get up into town, such a long way from the docks that the trip had to be made by horse and buggy or one of the few old taxi cabs available. There wasn't much to see in San Salvador other than the bars and the usual houses of ill repute. We drank some beer, which tasted pretty good after the long trip south, then it was back to the ship and on to Panama to offload more cargo.

From Panama we went to Guayaquil in Ecuador. A lot of the sacked grain loaded at Portland and Seattle was discharged there. We were allowed ashore to visit the city, warned not to get in trouble or we'd be kept in jail there more or less permanently. The final stop after Guayaquil was in Peru, a small port north of Callao where we discharged into lighters and were never able to get into the town. From there we went back to La Union to discharge some goods not unloaded on the first stop and then to San Marcos Island to take on another load of gypsum, this time bound for Seattle.

Arriving in Seattle in March of 1926, I decided to leave the ship at that point. Sailing on the *S.A.Perkins* had been a wonderful experience for me. A good crew and deck officers, excellent food, smooth sailing for the most part except for an occasional storm in the Gulf of Tehuantepec off southern Mexico, all had combined to remove any remaining doubts I might have had about a lifelong career at sea. But I was young and had itchy feet. My friend Jim Murray left the ship at the same time. It was such a good ship that the rest of the crew remained together aboard her for a long time after that.

I was home for awhile on Bainbridge Island, then moved over to stay at the Stevens Hotel in Seattle as a base for job hunting. Jim and I linked up at the hotel with another friend, Kenneth Voight, also looking for work on a ship. One morning Jim came back to the hotel to tell me that we had a job. The ship was the *Waukena*, an old shovel-nosed, twin screw Puget Sound freighter that had been laid up for a few years and was being recommissioned by the Borderline

Transportation Company. Jim was hired on as quartermaster, while I was on the deck gang.

After the first stretch of work, I wished I was back on the *S.A.Perkins*. We worked thirty-six hours straight, loading cargo at various docks around Seattle, then had only a few hours rest before arriving at Everett to discharge some cargo there. We did get more rest on the way to Bellingham to unload the remaining cargo, after which we ran to Lime Kiln on San Juan Island. There we worked and worked loading lime, all of us in the deck crew growling about the hard work and poor food. The captain made it known that we'd find ourselves walking to Friday Harbor on the island if we kept that up, so we went on working despite the conditions. By the time we got back to Seattle we'd been out about ten days or two weeks, enough for me. At the $60 a month rate, I got all of about $30 for the work done during that time.

It would soon be springtime, so I went back to Bainbridge Island to help Dad paint *La Blanca* and generally get her ready for the summer berry season. Again I stayed with him through the summer, hauling berries, bringing fertilizer back from Seattle, hauling dynamite for the Du Pont Company, all the usual activities of Dad's business. That summer was another busy one, Dad chartering the *Saint Michael* and leaving me to run *La Blanca*.

All went well until mid-July, when I was leaving Eagle Harbor on the evening run with a barge-load of strawberries. Just as I got off Parker's Landing there was a terrible noise from below and the engine stopped. I ran down to the engine room to check the trouble, finding that the crankshaft had broken on the old Eastern Standard engine that had served so well for eighteen years. There was nothing to do but get a message to Dad, who sent out a small tug owned by a Winslow friend, Joe Lundgren. Joe towed us back to anchorage at the head of the bay and then hauled the barge to Seattle.

A survey of the engine showed clearly that it was time for a change. Dad made a good decision at that point. He went to the expense of installing a new diesel engine in *La Blanca*, a three-cylinder Cummins that was one of the first of the Cummins line to be used on Puget Sound. Despite the expense, he won out in the long run because the diesel engine used only about three gallons of fuel an hour compared to the four or five gallons of distillate used by the old Eastern Standard, and the diesel fuel cost just six cents a gallon. At the time the old engine quit we were paying about eighteen cents a gallon for distillate.

Dad chartered a second boat during the several weeks it took to replace the old engine, continuing operations as usual until the berry season ended, when once again I was left to wonder what would come next.

The first opportunity was a ship that had come through the Panama Canal while I was there on the *S.A. Perkins*. She was a World War I surplus ship, the

City of Lordsburg, laying in an East Coast port when the Griffiths Company bought her. Re-named the *James Griffiths*, she was picked up on the East Coast with Captain Jack Clark as master. He took her to Texas to load sulfur, then through the Panama Canal and up to Seattle. Captain Earl Mercer from the *S.A.Perkins* was the senior captain for the company at that time, so he was named as master of this, the newest ship then owned by Griffiths. Captain Clark remained aboard for a time as Chief Mate, then became master of a sister ship to the *S.A.Perkins*, the *Delight*.

I visited the *James Griffiths* when she called at Seattle, finding that the mate who replaced Jack Clark was an old friend of mine from Bainbridge Island, Mike Soderlund. It happened that the bosun had just quit, giving me a chance at my first job as petty officer. The bosun on a ship was rated p.o., as it was called, because he was in charge of the deck gang. His duties were to check with the chief mate as he was coming off watch at 8:00 A.M. to learn what work was to be done that day, passing the orders on to the AB's in the deck crew and supervising their work. I was awfully proud of myself when Mike gave me that job. At the age of twenty-one I was much younger than most of the AB's in the crew, but at the same time already had more experience than many of them.

The *James Griffiths* was in coastwise trade at the time. We hauled grain to the Sperry mill at Vallejo, near the mouth of the Sacramento River; brought general cargo back to Puget Sound, sometimes carrying lumber to San Pedro and often alternating cargoes between various West Coast ports. One run the ship made quite often was to Britannia Beach, in British Columbia, where we loaded copper concentrate from a mining operation of the owners' parent company, James Griffiths and Sons. The concentrate was brought back to Puget Sound for discharging at the ASARCO smelter in Tacoma, or taken to the upper San Francisco Bay area for delivery to a smelter at Selby, a few miles above Vallejo on the Sacramento River. One of the cargoes out of the Sound was flue dust from the smelter at Tacoma, a heavy material that was taken to the Selby smelter to refine further for gold and other residuals. They were also able to get more copper from the dust, which was poured into bars that we would in turn carry back to Tacoma.

It was during one of those runs to Selby with flue dust that I first experienced a man going overboard. We had discharged the dust at Selby during the day, working until nightfall. The crew was dirty from working with that dust. One of the sailors had to use the toilet, but found the head crowded with all the other sailors cleaning up from the work of the day. In a hurry, he ran to the stern of the ship, squatting over the stern while he hung on to one of the cables in the rail around the fantail. The cable gave way suddenly and over he went, yelling at the top of his lungs.

The ship was already under way when that happened, headed down the Sacramento River bound for Oakland. Hearing the sailor yell, we notified the bridge right away. The ship was stopped, turned around in the river, and we started searching for the lost sailor. It was no use. In the dark, with the river running strong, there was just no telling where he might be. We gave up after a couple of hours and turned back downstream to run to Oakland.

The captain was pretty dejected about losing a man overboard as we approached Oakland. It's easy to imagine how startled he was when, as we neared the dock, we found our lost sailor standing there with a grin on his face! A Red Stack tug following us down the river had heard him yelling from the water and picked him up. They passed our ship while we were searching for him, and so the sailor beat us to our destination.

The *James Griffiths* was an interesting ship to serve on because of all the different cargoes we carried and the variety of ports visited. As bosun I was making $75 a month, which I thought was good pay, and had my own stateroom aboard. The chief mate was a help getting me started in my first bosun's job, advising me the best ways to supervise the deck gang and how to organize the work.

There was always a lot of work to be done on deck. The deck crew's duties included all stages of painting, beginning with chipping and scraping rusty areas, painting the bare metal with red lead to retard rust, and finally painting the finishing coats. Paintwork in good condition was washed frequently, "soogying" as we called it, using a washing powder popular in those days called "Gold Dust." It came in a box with the picture of a pair of young black twins. I suppose that is where the expression "gold dust twins" originated.

One of the dirtier, less interesting jobs was scraping up and washing the decks where caked oil accumulated under cargo winches. A good bosun always gave such dirty jobs to the deckhands who were laggards or complainers. Most of them would get the idea and shape up after this had happened to them a few times. More interesting was working with manila and wire rope, putting new eye splices in various lines, renewing tackle on the lifeboats, etc.

There was a lot of variation in cargo handling. On coastwise ships it was quite often loaded and discharged by the ship's crew, but that depended both on the type of cargo and the ports where it was handled. We worked cargo as part of our job at this time, though overtime was given if the cargo work went beyond the normal watch, eight hours in the case of the *James Griffiths*. After the seamen's strike of 1934, extra pay was given whenever any cargo work was done by the crew, whether or not it was part of their regular work day.

Most of us in the deck crew of the *James Griffiths* thought we had good jobs and were treated well. Compared to many other ships we did have it good, a fact

I was to learn on some of my later shipboard jobs. We were aware, though, of some very poor treatment given the crews on other ships, and this awareness led to my joining the Sailors Union of the Pacific that same year. My brother Paul had also gone to sea about this time and he, too, joined that union in 1926.

I stayed on that ship until Christmas, when my folks wanted us all home for the holiday. I was paid off the *James Griffiths*, spent the holiday with the family, then went looking for another job. Going to the Sailors Union proved useless, as with few exceptions the companies would not call in with their available jobs. We would hang around the old Grand Trunk Dock office of the United States Shipping Board, forerunner of the Maritime Commission, and at the Pacific Steamship dock where the shipping master for the company, "Con" Campion, ran a hiring place for the line. Pacific Steamship was also known as the Admiral Line because of the large number of company ships bearing the names of admirals.

One day while I was at the Shipping Board office I learned that a steamer was going to be brought out of mothballs to go on a run for the Swain and Hoight Shipping Company. The ship's name was *Cross Keys*, built locally by the Skinner and Eddy shipyard during World War I. She was due to load for South Pacific ports, Burma, Java and others, and was hiring a crew. I got on the crew as an AB in January 1927.

We took on a general cargo at various points around Puget Sound, went down to Aberdeen for more of the same, then to Eureka for lumber and finally back to Aberdeen to top off the load. Arriving off Aberdeen that second time there was a terrible blow, forcing us to stay at sea off the coast until the weather subsided so we could cross the bar into Grays Harbor. When the weather started to moderate we ran down to the Columbia River for a pilot to take us across the bar at Grays Harbor. It was quite a sight that day, as many ships from all over the world had been standing off at sea, waiting to get into the Grays Harbor port facilities at Hoquiam and Aberdeen. After we got back to the area with the pilot, the first ship crossed the bar and then there was a long line of ships formed, running across the bar one by one.

We made it into Aberdeen and were finishing off the load, the weather still nasty with wind and rain. While we were working cargo some young fellows had foolishly taken a sailboat out in the harbor in those conditions, overturning it and screaming for help. Our bosun, a witty Irishman named Campbell, had us lower our ship's boat to rescue them, the sailboaters all the time keeping up their screaming. Finally he yelled back at them in his Irish brogue: "Ah, shut up, ye got fair wind!" We got the boat over the side and managed to rescue the boys, which was the highlight of an otherwise not very happy time.

The *Cross Keys* was a disappointment to many of us on the crew. The food was both poor and stingy. A few cans of condensed milk were doled out for mush at breakfast, never enough to go around, and the mess was locked up tight except for three meals a day. No midnight lunch was provided, nor could one even get a cup of coffee outside those meal hours. Adding to that was an unpleasant captain, so several of us decided to get off the ship before she left Aberdeen.

I found things pretty tough in the shipping business when I got back to Seattle in March. Nothing turned up at the usual offices where we went looking for jobs. Finally I ran into my old friend the chief mate on the *James Griffiths*, Mike Soderlund, who had been promoted to captain. Griffiths Steamship had bought three more ships known as "lakers," with four hatches forward and engines aft, which were good for use in coastwise trade. When Captain Soderlund learned I was out of work he offered me a job as an AB on one of these ships, the *El Cedro*, then loading for Mexico at Pier 6 in Seattle. Though licensed as a captain, Mike Soderlund was sailing as chief mate under another company man, Captain Joyce. I was delighted to have a job again and to be sailing with Mike. I was even more pleased to learn that our first destination was Guaymas, Mexico, where I'd have a chance to renew acquaintances.

On the way down the coast we went into San Pedro to take on fuel. It had become standard practice for these ships to fuel up there or at Los Angeles, since both were close to oil fields that produced a good quality of crude oil. As with most of their other vessels, the Griffiths Company had put a big towing winch on this ship as soon as they bought her. On leaving San Pedro we picked up a tow for Mexico, a big barge named *Daylight*. She had been a four-masted bark, rigged down with the top masts removed when Griffiths bought her to use as a barge. It reminded me of my time on the *Griffson* as we towed her to San Marcos Island for a load of gypsum, then went on with *El Cedro* to Guaymas. During the ten days we lay discharging cargo at that port I had a wonderful time visiting my old friends from *Griffson* days.

When the cargo was offloaded we ran back to San Marcos Island and took on our own load of gypsum, the usual dirty job with dust and grit flying all over the place. While I was working with the loading somebody came down to tell me the captain wanted to see me in his quarters. Dirty and grimy as I was I went to see him. "You sent for me, Captain?", I asked. "Yes, Christensen," he said, "I want you to go get cleaned up and relieve the messman for the officers' mess."

I was shocked. "Why me, Captain? I'm an AB, no messman."

That didn't have much effect on the old man. "The dining room messman went ashore while we were in Guaymas and came back with a dose of the clap. I don't know what else he brought back with him, but I know I don't want him waiting on tables in our mess. I've been observing you and you look like a pretty clean-cut young fellow. I've talked it over with the other officers and they want you waiting on tables."

"Oh, God," I said. "I don't want to do that. I hired on as an able-bodied seaman. I'd rather be out working in the muck and the dirt than waiting on tables and having to make up the officers' rooms."

He still wasn't going to change his mind. "Why don't you try it for a day or so? We'll see how it goes."

"I can't refuse you, Captain," I said, "if you want me to do it. But I'm doing it reluctantly." I got cleaned up, went to the mess, set the tables and took the orders. Next morning I got up early and did the same thing over. I was mad as hell. I didn't want to be a messman. It was a terrible comedown for an able-bodied seaman.

I stuck with it through most of that day, cleaning the officers' rooms and making up the bunks. The engineers' rooms were terrible, dirty and oily, and finally I thought "to hell with it. I don't want this job. I'm going to go up to tell the captain, and if he still wants me to keep this job I'll quit, even if he kicks me ashore."

When I went to see the captain I told him I couldn't handle that job. "You have a son aboard this ship who's a nice clean-cut young fellow," I told him. "Why didn't you offer the job to him?"

"I haven't asked him, and I don't know if he'd want to take it."

"Well, ask him. They're going to have trouble with me if I have to keep on waiting tables and cleaning rooms. You have authority over your son."

Captain Joyce finally saw it my way. He called his son up to his quarters and told him that I would rather be mucking around in the grime and dirt than doing messman's duty. He ended up talking his son into taking the job. I was happy as a clam to get back to deck work again.

When the loading of gypsum was finished we hooked on to the *Daylight* and towed her up to Long Beach. Dropping her there, we then ran into San Pedro to take on fuel. *El Cedro* was set up so the number two hatch could be used as an oil tank, and she also had wing tanks that could be used to carry a cargo of fuel. While taking on fuel for the ship's engine we filled these tanks with a load of crude oil for delivery to the Union Oil Company at Edmonds, on Puget Sound, about 15,000 barrels in all. Carrying this cargo from San Pedro to Puget Sound earned enough money for the company to pay for all the fuel used by *El Cedro* on the round trip down the coast and back.

After arriving at Puget Sound and discharging the oil, we repeated the entire operation on another trip down the coast, again picking up the *Daylight* on the way down the coast, dropping her at San Marcos Island to be loaded with gypsum while we went to Guaymas to offload lumber and pilings. The trip back north was another repeat. When we arrived at Puget Sound on that trip it was June of 1927. Berry season again. I signed off *El Cedro* and went back to work for Dad.

My brother Paul was with us that season. When it ended, he and I were both in need of work and were lucky to sign on together aboard the *James Griffiths* as AB's. From the fall of 1927 to June 1928 we worked aboard her in the same activities as during my first time on her crew, running coastal trade between Puget Sound and the Bay Area and hauling copper concentrate from British Columbia.

It was a good year. A compatible crew with good officers in that kind of trade made for a pleasant life. There were some real characters in the crew. One was a fellow named either Paul Bruno or Bruno Paul. We never knew for sure which was his first and which his last name, as he didn't know himself. He was a big, nice fellow of undetermined origin who wasn't very smart but was a good worker.

Another character was Karl Orris, a White Russian who had escaped during the Bolshevik revolution of 1917, getting out of Russia through Turkey with many adventures. Karl was not as tall as Paul Bruno, but was a stocky fellow and a strong deckhand.

One time when we were unloading flue dust at Selby, Karl and Paul were working together in the hold. For reasons nobody ever learned they got into a terrible fight down there in the hold, which wound up with Karl hitting Paul so hard that he split Paul's nose wide open. Paul went running up to Captain Mercer to tell him what had happened. Never able to express himself very well, and with the excitement of the fight and his injury, all he could do was hold his hand to his bleeding nose and repeat over and over: "Orris, Orris, Orris." The captain was so concerned about the injury that he sent Paul up to a hospital to have his nose sewed back together.

Later that day, Captain Mercer called some of us from the deck crew to get the details of Paul's injury for a written report. When he learned it was the result of a fight with Karl Orris he was very angry. "For God's sake," he said, "I thought Paul was trying to tell me he had been hit by the ore bucket. If I'd known his nose was split in a fight I'd never have wasted the company's money to get it fixed!"

Our ship called at Oakland and San Francisco quite often to load or discharge general cargo, giving us time to get to know those cities. Wherever we docked there were vendors trying to sell their wares to seamen. In San Francisco, one of the people who met the boats was a clothing salesman, Fred Harris, who had a store with a good line of top quality uniforms and clothing for seamen. Another was Sam Harris, a shyster who owned a similar store that carried inferior clothing at inflated prices. Sam Harris spent a lot of time soliciting aboard the ships, and implied that he was Fred Harris in order to ride on the better reputation. It got so bad that Fred Harris finally started passing out cards with his store name and address that had "No relation to Sam Harris" printed at the bottom.

By now it had become standard practice for me to sign off whatever ship I was aboard in June, in order to go back with Dad to help him during the busy summer. Dad relied on me for this, but it was no hardship as he paid me well and I truly enjoyed the activities and camaraderie of that work. So I did that once again in June 1928, this time staying with Dad only through the strawberry season. When that ended, about mid-July, Dad didn't need my help any longer. While I was looking for work it happened that Dad ran into an old friend in Seattle, Captain Odson of the Alaska Steamship Company. He was a fine, fine gentleman, a slow-talking older Norwegian who had been with the company a long time. The captain told Dad that he was taking one of the company's ships out of the "bone yard," as we called it in those days when a ship was laid up for a time and then put back into service, in this case the *Oduna*. Dad mentioned that I was looking for work and suggested that Captain Odson give me a call if he had an opening on the deck crew.

The captain did call me with the offer of an AB's position, which I gladly took. We loaded general cargo for the north before our departure, then stopped at Nanaimo in British Columbia to take on a load of coal destined for Cordova. From there we headed north, stopping at many small ports in Southeast Alaska to discharge cargo and pick up canned salmon for the return trip. After the coal was unloaded at Cordova we took on a lot of worn out machinery from the Kennicot mine, including a locomotive loaded on deck. We also went to Latouche, in Prince William Sound, to take on a deck load of heavy machinery bound south, and some copper concentrate ore destined for the ASARCO smelter at Tacoma.

The deck crews worked all the cargo on these runs because of the small ports with few facilities, so I was pretty tired when I took wheel watch as we left Cordova around midnight sometime in the second week of August 1928. After I had taken the wheel, Captain Odson asked in his slow-talking way: "Holger, did you hear about your father?"

"No, I haven't, Captain," I said, "I've been working the clock around loading the ship. I haven't heard anything."

"Well," he said, dragging the words out slowly, "the boat blew up. Nobody was hurt. I have the *Seattle Star* newspaper down in my cabin. When you get off watch you can read all about it." When I came off watch at 4:00 AM I got the paper and read that *La Blanca* was no more. I felt awful about it, though very happy that nobody was hurt.

In later years, I might have flown back home to help Dad out in his distress with the loss of the boat, but there was no such possibility in 1928. I stuck out the rest of the trip on the *Oduna*, arriving back in Puget Sound in early September 1928. The ship had been returned to service for just this one trip, so she was due to go back into Alaska Steam's West Seattle bone yard. Before putting her in there, though, we had to deliver the various cargoes picked up in Alaska. We stopped first in Seattle, where Captain Odson picked up his son to be with him as we shifted to Tacoma to discharge the ore. The captain had Paul, then about twelve years old, steer the vessel on the way down to Tacoma, asking me to stand behind the boy to check that he steered all right. I didn't have to do anything in the way of correcting him, as he was already a skilled helmsman at that young age. Following in his father's footsteps, he was also to become a maritime captain.

After discharging the final cargo in Tacoma we returned to Seattle to put the ship back into lay-up at the company's West Seattle yard, arriving there about midnight. Besides some finger piers at the yard there was one parallel to the waterfront, where a steamer lay moored. She was the old *Latouche*, a small cargo ship with the engines aft and cargo holds forward. I was standing on the after deck, ready with mooring lines, as we approached the yard. Suddenly I could see us picking up speed.

It was one of those rare instances when the signals between the bridge and engine room got mixed up. The captain had rung for full astern on the engines, while the engineer on watch had misunderstood and put them full ahead. The bridge put the engine room telegraph at stop, then rang for full astern again, but it was too late. The ship had picked up too much speed.

The chief mate, a big Finnish fellow, was standing on the focsle head. When he saw what was happening, he let go both anchors in an attempt to check the ship's speed, ordering the crew to put the brakes on the windlasses as soon as the anchors took hold on the bottom. It slowed the *Oduna*, but not enough. She hit the *Latouche* so hard that our bow went right through her side at the engine room, plowing about half way through her all the way up to the smoke stack. It was fortunate nobody was aboard the *Latouche* and that none of our crew was hurt in the accident. The *Oduna* had her bow bent up, damage that was later repaired at that same yard, and there was a hearing which determined that the engineer was to blame.

After we got disengaged from the *Latouche* and got the *Oduna* properly moored, I packed my bags. It was time to get home to learn more about the loss of *La Blanca*, and my father's plans now that his principle means of livelihood was gone.

Sea Travels

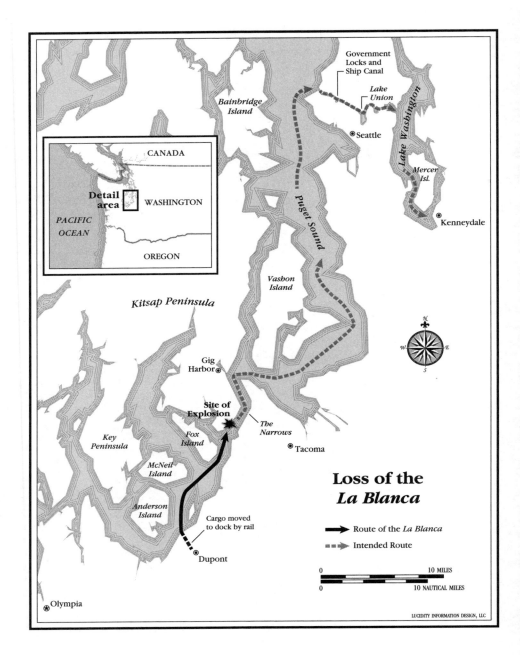

PUGET SOUND

CHAPTER EIGHT

LOSS OF *LA BLANCA* BUILDING THE *HANNAH C.*

THE *HANNAH C.*

The blast that destroyed La Blanca *was, to the best of my knowledge, the second largest accidental explosion ever to be recorded on Puget Sound, exceeded only by the 1918 explosion of 90 tons of dynamite on a barge moored in Elliot Bay, off Seattle. It was "only" 12 tons of explosives that blew* La Blanca *apart, but that much black powder and dynamite had quite an effect. Although I wasn't there, by going ahead in my story to World War II days I can give an idea of how impressive that explosion was.*

Sea Travels

IN APRIL 1942, I was assigned as second mate to a brand new Liberty-class ship delivered for the Griffiths Steamship Company by the Kaiser Shipyard in Portland, Oregon. Under command of Griffiths' Captain Eckert, our crew took delivery of the *Jonathan Edwards*, made a short shakedown cruise, then proceeded to San Francisco to load general cargo bound for Pearl Harbor. As usual during those wartime years, we crossed the Pacific to the Hawaiian Islands in a convoy with many other merchant ships.

After discharging at Pearl Harbor the ship went on to the port of Kahului on the island of Maui, where we loaded 11,000 tons of sugar. I still find it interesting that the destination of this cargo was kept secret from the crew, while Japanese longshoremen told us they knew it was bound for New York because of the way it was loaded.

Next, we made the 4,600 mile crossing of the Pacific to Panama by ourselves. It was not a very comfortable feeling to be alone on the Pacific Ocean at that time, but we made it without incident and passed through the canal to wait in Cristobal for a convoy to be made up, the first convoy of the war through the Caribbean. Lone merchantmen were starting to be sunk by German submarines operating in the area, so from this time on throughout World War II all merchant ships were required to travel in convoy.

After some 50 ships had gathered we headed north with two destroyer escorts and air cover, taking two days and one night to arrive at Guantanamo Bay, Cuba, where we anchored up for the night. As we left Guantanamo the following morning, other convoys gathered with ours, ships in such numbers that the entire sea seemed to be covered with them. Danger from German submarines was so great along the coast north from here that we would be traveling in daylight hours only from now on.

There was another night layover at Key West. After departure the next morning my first bridge watch was from noon to 4:00 PM. By now it was May of 1942. The weather was slightly overcast, visibility was good, the sea smooth with a slight swell. The submarines struck suddenly in an attack that lasted about a half hour and was most apparent to me from the U.S. Navy escorts running at high speed while dropping depth charges, and the explosions and smoke from two ships sunk near the tail end of the convoy. The *Jonathan Edwards* was not attacked.

We ran in to Chesapeake Bay in the early evening, again to lay up for the night because of the dangers of submarine attacks. When I went down to the dining saloon for dinner I was seated with the ship's second engineer. He was still shaken by the afternoon's action.

"It scared hell out of me," he said, "to be down in the engine room during the attack. The depth charges make a terrible noise when you're down under the water line on a steel ship. The only thing I can compare it to was when I was on the *Margaret Dollar* back in 1928. We were steaming south through the Tacoma Narrows on Puget Sound when a powder boat blew up near us. I was down in the engine room when that happened, and that scared hell out of me, too."

We were both amazed by the coincidence because that powder boat was, of course, *La Blanca*.

* * *

LA BLANCA's last voyage began uneventfully as had all the trips through the twenty years she had been in service. Dad and my youngest brother, Nels, had run the boat down to Du Pont on the morning of Monday, August 6, 1928, to load dynamite and black powder for a development project at Kennydale, located at the southeast end of Lake Washington. To reach Kennydale they would pass through the Government Locks which separate Puget Sound from Lake Union, nestled in the heart of Seattle, and then eastward through Portage Bay into Lake Washington.

On that Monday they loaded 260 cases of dynamite, 225 kegs of black powder and some boxes of detonating caps. In my mind's eye I can still see the little town of Du Pont, sixteen miles south of Tacoma, the plant dock where we tied up to load, and the loading process itself. Part of that memory goes back to the time when I was a young boy riding along with Dad, and part as I matured to making those runs as operator of *La Blanca* with my next-youngest brother, Paul, as my crew member.

The Du Pont plant was located about two miles back from the water. Explosives to be shipped by water were taken to the dock on a small narrow-gauge railway built by Du Pont, in boxcars pulled by a diesel locomotive. Once at the dock, the materials were sent down to the deck of the boat in wooden chutes, which had flexible oak slats that could be adjusted to act as brakes depending on the stage of the tide and difference in height between the dock and the boat.

Black powder was packed in small metal kegs, each weighing twenty-five pounds. Our normal practice was to stow these kegs in the hold, just forward of the engine room bulkhead, fore-and-aft on their sides because they fit the natural curvature of the side of the boat this way and were secure from rolling around in rough water. Cases of dynamite were generally loaded on deck, though we occasionally mixed dynamite and black powder in the hold depending on the size of the load and the destinations. There was no regulation against this mixing of powder and dynamite, but there was a regulation concerning blasting caps. They were packed in small tins, which in turn were packed into wooden boxes with sawdust to minimize shock. The regulation required that they be stored separately from the powder and dynamite. On *La Blanca* we stored them in the lazarette, a small hold at the aft end of the boat.

By all accounts Dad and Nels loaded *La Blanca* that day just as I have described, with the black powder kegs in the hold and dynamite on deck. It was 5:20 PM when Nels threw off the mooring lines and Dad backed *La Blanca* out into the stream, then went into forward gear with the pilot house controls and swung north to pass through the Tacoma Narrows on the way towards Seattle. They had been under way for two hours when they reached a place one-half mile south of where the Tacoma Narrows Bridge stands today, some eight miles west of downtown Tacoma. Dad was in the wheelhouse, while Nels was preparing supper down forward in the galley. When Dad looked aft to make one of his frequent checks of the cargo, he saw smoke coming from the tarpaulins covering the cases of dynamite on deck. He yelled for Nels to come up from the galley, and together they ran back to check on the smoke.

By the time they reached the after deck the tarpaulins had burst into flame. They immediately tried to put out the fire, but it was quickly apparent that they could not extinguish it. The dynamite itself was already starting to burn. There was no danger of explosion from that alone, but when the fire reached the black powder in the hold the boat was doomed. Their temptation was to abandon ship right there, a temptation that Dad put aside at once. The *Margaret Dollar*, a big steamer, was steaming towards them only a half mile to the north in the narrows, and the good summer weather had brought out many small recreational boats. Some of these boats rushed to the assistance of *La Blanca* while Dad, setting a course for the beach on Gig Harbor Peninsula, yelled at them: "Keep off. Dynamite. She's got to blow!"

Thankfully the small boats turned and ran. When *La Blanca* hit the beach, Dad and Nels scrambled off and waded through the waist-deep water of high tide to take shelter in a little valley off the beach, surrounded by trees. In a very short time, just ten minutes after the fire was discovered, there was a loud hiss, a flash, and then the huge explosion.

The blast broke countless windows in Tacoma and caused atmospheric concussion felt as far north as Seattle, 40 miles away. Beached at the base of a 70-foot cliff, *La Blanca* simply disappeared except for the pilot house, which flew 100 feet in the air to land on the rear slope of the cliff, where it started three forest fires that quickly spread into one. A crater ten feet deep was left in the beach gravel where the boat had been, and other than the burning pilot house only a few charred embers floating in the water had survived the blast.

Many of the above details are from an article appearing the following day in the *Seattle Daily Times*, which reported that my father and brother were picked up after the explosion by a young man in a row boat, then transferred to a speedboat that took them to Tacoma. The *Times* quoted Dad as saying: "We fought it as long as we could but I could see it would reach the blasting powder any minute, so Nels continued to fight it back while I ran inshore. I can't say much about it until I talk with the Du Pont Company officials. They don't like to have things like this get into the papers and it might hurt me if I get another boat and want more powder freighting contracts." Asked if he really would continue to carry powder after this experience, Dad replied: "Sure, why not? These things only happen once in a hundred years. I'll probably never have another explosion aboard but if Nels is with me we'll get by just like this time."

The *Times* treated Dad and Nels kindly in this article published the day after the explosion, saying: "Only their heroism in running their blazing launch and its dangerous freight out of the shipping lane prevented loss of life, which must have occurred had they continued on their course. They would have passed the liner *Margaret Dollar* and a school of small craft that had started to their rescue."

But a hint of trouble to come appeared in the same article in a quote from Mrs. R.H. Calkins, whose home on the cliff over the grounded boat was completely destroyed. She and her husband had been fishing in the stream about 100 yards from *La Blanca* when the boat caught fire. Hearing Dad's warning, they had run out from the beach. Mrs. Calkins said of the experience: "It was terrible. That awful hiss while we waited for the explosion. And when it came it knocked me to my knees in the bottom of our boat. Our home that we have worked on for two years is gone. There's not even a thing to eat left."

In its issue the day following this article, the *Times* editorialized in a vein quite different from the tone of the article praising the heroism of the crew of *La Blanca*. In bold print on the front page of the *Times* edition of August 8, 1928, was this editorial:

SAFEGUARDS MUST BE PROVIDED IN TRANSPORTING EXPLOSIVES

"To carry black powder and dynamite in the same craft or in any other kinds of conveyance is contrary to every rule of common sense. It is in direct defiance of natural law inherent in the character of these two explosives. It is strictly prohibited in Army and Navy practice; and in any circumstances, anywhere, at any time, is wholly inexcusable.

"This fact is enough in itself to detract much from the 'heroism' of the two men who, having loaded their flimsy boat with black powder and dynamite, ran it ashore when fire broke out in the powder while passing through the Narrows near Tacoma last Monday. That the two men were able to make a landing and get away before the

inevitable explosion occurred was a piece of good luck for them. That it all happened at a point where there was no one else to be endangered nor much property to be damaged, was simply a providential diversion of the almost certain consequences of foolhardiness.

"*This highly dangerous cargo was destined for a point on Lake Washington, to reach which it would have had to pass through the government canal. An explosion at any point along that route would have wrought havoc beyond any present power of imagination. Death and destruction would have been spread over large and thickly populated areas of Seattle. There was no particular reason why the black powder should have caught fire while the craft was in the Narrows. The blaze might just as easily have started while it was being lifted in the canal locks at Ballard, with results too fearful to be appraised.*

"*Carelessness of life and property is at the bottom of such doings, to be sure. But where lies the responsibility for the lack of law, federal or state, the failure of regulation to check against such carelessness and to prevent the incurrence of such terrific risks? Somewhere, and with someone other than the two foolish men who were lucky enough to escape with their lives, that grave responsibility rests.*"

That editorial was bad news for Dad. He had, in fact, followed regulations throughout the many years *La Blanca* had hauled explosives for Du Pont, regulations that included the loading procedures I've described, flying a red flag on the boat to indicate the carrying of explosives, and arranging a permit with Seattle's city government for carrying the explosives through the locks to Lake Washington. The exact nature of the cargo was furnished to obtain the permit, which was valid for daylight hours only.

What made the editorial bad for my father was that it invited law suits, which were not long in coming. The suits came from insurance companies, trying to recover their losses in paying off on policies covering losses of homes and businesses damaged by the blast. Both Dad and Du Pont were named in the suit, a fortunate thing for Dad because the huge Du Pont Company could afford to hire the best lawyers to defend against the suits. The cases dragged on for a long time before finally being thrown out of court by a Tacoma judge who ruled that nobody in his right mind would set fire to his own boat loaded with explosives, so the fire and explosion had to be "an act of God."

Nobody ever did determine what caused the fire. The most probable cause, though, was that an ember from forest fires burning on the Olympic Peninsula had alighted on the tarpaulins and set fire to them. It was a hot August day with the wind blowing offshore from the peninsula, and Dad had been hugging the beach to find the slackest water while running against an incoming tide, all of which would suggest that the forest fire theory is the right one.

Dad was in quite a fix without a boat to carry on his business, but he was undaunted. By the time I arrived home from that trip on the *Oduna* he was already making arrangements to lay the keel for a new boat. Financing had been a problem, he told me, mainly in getting a co-signer for a bank note to cover about half the cost of construction. He had savings of about $10,000, enough to cover building the kind of boat he wanted. But he needed that much again for a proper diesel engine, and the bank insisted on a co-signer for that much money. He had first gone to the president of the National Fruit Canning Company in Seattle, a Mr. McCaffery, with whom Dad had dealt honorably for years in connection with the berry-hauling business. For some reason Mr. McCaffery had turned him down, so he then went to R.D. Bodle, owner of the canning company by that name and again a person Dad had dealt with for years. Mr. Bodle did co-sign the note, plans had been drawn up, and the keel was about to be laid at the Olson and Sundy yard located at the foot of 6th Avenue N.W. in Ballard.

Construction had just begun when I first visited the yard. My brother, Paul, had also come home from deep-water sailing to be with Dad during this period. We decided to give up our careers at sea for as long as it would take to help our father get back on his feet with the business, helping out around the shipyard as she was built. It took less than three months to complete the hull and superstructure. After she was launched the engine was installed, a 100 horsepower Atlas marine diesel with three cylinders. I remember our watching as the ten-ton engine was lowered onto its mounts. The vessel hardly went down in the water at all with that much weight, which pleased Dad immensely. "Now I can see," he said, "that I'll be able to carry a lot of freight on her."

She was a nice boat, the *Hannah C*. Built along the lines of an Alaskan fish tender she was sixty-five feet in length, constructed on heavy oak frames with true 2-1/4 inch fir planking. The engine was towards the aft end, with a small cargo hatch behind that. The pilothouse, galley and a small bunkroom were also aft, while the main cargo hold was forward of the pilothouse. The focsle was forward with a raised focsle head, accomodating six bunks. There was a hefty towing bit on the after deck. Altogether, she fit the needs of Dad's business as well as any boat could.

Paul and I helped Dad paint and equip the *Hannah C.*, rigging, fitting her out with lines and all the equipment necessary to get back into business. It must have been about late December 1928 when we were finally ready to go back to work. It's ironic that the very first load carried aboard the *Hannah C.* was from

Du Pont, sixty-five tons of dynamite for transshipment to a large steamer anchored off Blake Island near Seattle. We picked up the dynamite at the Du Pont plant, brought it back to Eagle Harbor where we anchored for two or three days waiting for the steamer to get ready for departure, then ran out to discharge the load onto the ship. The Du Pont Company remained faithful to Dad despite the accidental explosion, giving him as much business as they could.

With a big debt load, Dad had to work the new boat hard to pay off the loan in the five years of the note. He got as many odd jobs as he could that winter, hauling freight of various kinds and getting towing jobs wherever possible. One contract was to haul cement for a riprap being built near Hoodsport. We made many runs on that contract.

A towing job I remember well was for the Roy Lillico Company, a small tugboat outfit. We went to Seabeck to tow three custom spars destined for England, each 150 feet long. Dad was hesitant to take on that job because Lillico wasn't very good about paying his bills, but he decided to take a chance because he needed the work. It wasn't a very successful job. For one thing we had trouble with the log patrols as we towed the spars up through Hoods Canal and then down to Seattle through Puget Sound. Log patrols scavenge beaches for useful timber, and for some reason the ones we met along the way got the idea that we were scavenging in their territory. They kept pestering us to turn the spars over to them. They certainly should have known that those very long, beautiful spars weren't salvage, so we just chased them off. When we arrived at the ship that would carry the spars, anchored off Smith Cove, they were hoisted out of the water and strapped to the side of the vessel, being too long to carry on deck. We got a receipt which was turned over to Roy Lillico, but he lived up to his reputation and never paid my father for that job.

The winter continued with odd jobs like those, while we continued to bring the *Hannah C.* into first class shape. Always pretty good with my hands, I built lockers and a workbench in the engine room. One evening we ran over to Seattle, planning to lay there that night to bring back sugar in the morning in preparation for the berry season. When we tied up about 7:00 PM, Dad reminded us that we had to get up early to pick up the load of sugar.

"We never replaced the alarm clock for the focsle we lost with *La Blanca*. Paul, I want you and Yankee to go up town and buy one." Dad was getting tired of having to come down to the focsle to wake up his crew. He gave Paul five dollars to buy the new alarm clock.

Paul and our friend Yankee McLaughlin, who was crewing with us on that trip, took off for town. They hadn't returned by the time Dad and I went to bed, and when Dad woke me in the morning I saw they still hadn't come back. Dad was grumbling about that as he and I fixed our breakfast and sat down to eat.

About that time a very tired-looking Paul and Yankee came into the cabin and sat down, Paul putting an alarm clock on the table. It was an awful looking thing, well-used, purple, with gold filigree around the edges. Dad looked at it sourly.

"God, that's an awful looking clock," he said, "Where did you guys get it?" Paul and Yankee sat there saying nothing.

"Looks to me," said Dad, "like that alarm clock was stolen from a whorehouse." The two culprits were still quiet.

Dad stared at them. "If I'm right, seems to me I've got a dollar change coming back."

Neither Paul nor Yankee ever said where they got that alarm clock. But Paul liked it well enough that he built a nice little wooden shelf for it in the focsle, where it was still sitting the last time I was on the *Hannah C.* many, many years later.

By the time berry season arrived the new boat was fixed up to my father's satisfaction. Working that season was much the same as in past years, with the differences being in the size of the operation and the size of the new boat. By now the berry operation on Bainbridge Island had grown to the point that as many as 500 barrels a day were filled with strawberries and sugar for transport to Seattle City Ice. Even with the increase to sixty-five tons per load aboard the new boat, Dad had to use barges to carry all the fresh and barreled berries.

The business of hauling blasting materials for Du Pont picked up about the same time, so Dad chartered a second boat to handle the berries while I ran the *Hannah C.* with Paul as my crewman, to carry dynamite and black powder. I remember the first of those runs came after we'd been working the clock around on the berry run. Paul and I were dog tired when we left Eagle Harbor, arriving at Du Pont about 4:00 in the morning and catching just a couple of hours sleep before we loaded dynamite and black powder for Port Angeles and Clallam Bay. Both of us were good cooks, but Paul was the better, so he fixed breakfast as soon as we had loaded and got under way. Then we kept running while taking turns at sleeping, about four hours on watch and four hours in a bunk just behind the wheelhouse. That was a handy location because if I was at the wheel and wanted to be relieved, all I had to do was open the door to the bunk room at the aft of the wheelhouse to call Paul.

We were pretty well organized to run that boat with just the two of us, but it was hard and tiring work. The same was true of the berry run, where it would take four or five extra men to handle all the loading and discharging of berries. The crews worked long hours, got little sleep and sometimes missed meals, but thinking back it's remarkable how little complaining was done. People in those days were glad for the work and enjoyed working together even if the job was physically tiring and had other hardships.

Somehow we always had fun working the berries. Everyone was jovial no matter how tired, there was lots of activity and many people involved. Quite often we'd carry extra people along with us as passengers, especially when going in through the locks to Lake Union. Going through the locks is an adventure for someone who has never done it, so we were always happy to take along some passenger who had never had the experience.

During the late summer of this year we started on a contract that was new for Dad, hauling canned salmon for Anthon Bugge. "Buggy", as he was known, was a big, happy, cigar-chewing Norwegian who had a cannery in Sequim, and a floating cannery that was then anchored at Neah Bay. From July to September we hauled empty cans to Sequim and Neah Bay, and the filled cans back from there to Seattle. When the season ended out at Neah Bay, near the point where the Strait of Juan de Fuca joins the Pacific Ocean, Buggy moved the floating cannery into the San Juan Islands for the late salmon season, anchoring it at Richardson on Lopez Island. Fishermen in the area would bring their salmon directly to the floating cannery, which was a large steamboat with living quarters aboard for the canning crew. She was named the *J.R. McDonald*, formerly used on the Columbia River and a fine barge for the kind of work Buggy did with her. When anchored at a place such as Richardson, local people also found work there but returned to their homes in the evening rather than living aboard. We continued hauling the empty cans and canned salmon for the rest of the season.

I recall that one time while we were laying overnight at Richardson, a couple of the island girls invited Paul and me to a dance. We thought that was something different; being invited by girls was quite unusual in those days. To make it even better they had a car, so we were driven to the dance and had a great time. When the dance was over the girls, who were sisters, decided they would drive up to show us their home at the north part of the island before taking us back to Richardson, on the south part. As the girl who was driving turned into their driveway and came to a stop, suddenly a hand came through the open window and grabbed her. It was the girls' mother. "You girls come out of that car and into the house!" she said. "You can't go out with those boys anymore."

That left Paul and me with no way to get back to the boat except by foot. It was a long hike, probably five or six miles, the first and only time I had to walk back from a car ride.

That was the kind of incident that got into the folklore of those small communities, one of many that people talked about for years. Another that didn't involve me but that I heard often was about an old fish buyer who came to stay aboard the floating cannery. They fed their crews well with meals prepared by lady cooks who also lived aboard. The first night this old fellow sat down to supper the gravy boat was already on the table. He grabbed it, ate the whole thing and said "My goodness that soup was good. Give us some more!"

The same fellow was watching one hot summer day as we were loading canned salmon. I was dressed only in a pair of dungarees and still getting so hot that I said the water looked good and I might just dive in. "I'll give you four bits if you do," he said. That was all I needed. I dove in, and when I climbed back aboard asked for my four bits. That was a lot of money then, so when he gave it to me I said "I'll take all my clothes off and dive in for five dollars." He backed off that one in a hurry.

At the end of the salmon season we hauled the floating cannery to Seattle, to be laid up for the winter in Lake Union for general maintenance and repairs to the machinery. Then we went back to work for Buggy on a different operation, canning clams at his Sequim cannery. He had several clam beds around the Sound, picking up the clams from those beds with his own boat, the *Phoenix*, about the same size as *La Blanca*. He couldn't keep his cannery fully supplied with clams, though, so he contacted a Canadian native group at Chemanis, British Columbia, where they had a large clamming operation. Dad was contracted to haul the clams down from Chemanis to Sequim.

That was quite an operation. At Chemanis, the clams were brought down to the boat in sacks, where they were loaded into boxes about the size of apple boxes, something to do with customs not allowing them to be imported in sacks. We would load twenty-five or thirty tons of clams, run south to Friday Harbor to clear customs there, then take the load on to the Sequim cannery. Everything worked well. Dad was smiling. Buggy was smiling and smoking his big cigar. Things worked out so well that we were even able to squeeze in some dynamite hauling for Du Pont occasionally.

Quite often we laid over at Friday Harbor after clearing customs, leaving early the next morning in order to get the clams to Sequim by the start of the working day at 8:00 AM. It was during one of those layovers that I again got myself into the folklore of the community.

I had gone ashore in the evening to relax a little, beginning with a visit to a gin mill close to the waterfront. My youngest brother, Nels, was along with us on this trip but was too young to go into the gin mill with me. After several drinks I was very relaxed, and when I left the place in that condition I found Nels waiting for me outside. He said he was going up to the drugstore to get a milkshake. That sounded like a good idea to me, so I went along.

As Nels and I sat at the counter in the drugstore, he with his milkshake and me with a cup of coffee, I saw a pretty girl sitting across from us. Being in my happy condition I started joshing with her, probably trying to make a date. She was a jovial girl, joshing me right back. We were just sitting there having a good time, the girl laughing a lot, when the druggist started taking offense. He told me to shut up, to which I gave him some sass, and suddenly he got so mad that he picked up a hammer and was going to hit me on the head with it. Lucky for me Nels was a big, strong football player, and sober. He took the hammer away from the druggist and got me to leave the place with him to go back to the boat.

In such a small community the story got around in a hurry. I took a lot of kidding about it, finding out that the druggist was sweet on the girl I had been giving a line. For his part, he got the nickname "Hammer Man," a common name for the men who ran the pile drivers so frequently seen around the islands in that time of development. That was a name that stuck with him for a long time.

The clam season came to an end just towards the end of the year. Paul and I spent Christmas and New Year at home. In January of 1930 he and I had spent well over a year helping Dad get back on his feet after the disaster with *La Blanca*. We felt good about that. Dad's new boat was well equipped and proven during a year's use, and it looked like he would continue to have plenty of business. It was time for us to move back into our own careers at sea.

CHAPTER NINE

Surviving the Great Depression

THE *JAMES GRIFFITHS*

IT WASN'T LONG into January of 1930 before I found a job, again with the Griffiths Steamship Company. They were taking *El Cedro* out of the boneyard for another trip, this one to the Aleutian Islands port of Dutch Harbor for the Northern Commercial Company. I was hired on as an AB.

Our main cargo was materials for a big dock under construction at Dutch Harbor, mostly heavy timbers and pilings. These were loaded around the Seattle area, the pilings brought aboard at the Coleman Creosote plant in West Seattle, with the number one hold left empty for general cargo that would be loaded in San Francisco. After loading that cargo at San Francisco we headed out across the Pacific direct for Unimak Pass, near Dutch Harbor. At that time of year it was a rough trip, with *El Cedro* laboring against heavy northerly winds and seas, about fifteen days from port to port.

The crew worked all the cargo on this trip. We offloaded the timbers and pilings at Dutch Harbor, ran to Unalaska to discharge general cargo for the Northern Commercial Company's operation there, then returned direct to Seattle in ballast, with no cargo.

That was to be my only deep sea trip that winter. It was April by the time we returned to Seattle and Dad was starting to get ready for the berry season, hauling fertilizer to Bainbridge Island for the farmers, barrels and sugar for the processors. Paul had also returned from a trip, so he and I once again went to work for Dad through the berry season, together running the *Hannah C.* both for hauling berries and for Du Pont when blasting materials were to be freighted.

One night towards the end of the berry season in mid-July, I had a night off and decided to go to the usual Saturday night dance at Fletcher Bay along with a friend, Erling Erickson. These dances were quite rowdy, but popular with the islanders. It was common to take a nip or two in preparation for the dance, which we did. When we arrived and took a look around I saw a pretty girl I wanted to dance with. I liked the cut of her jib. She was blonde, not much over five feet tall, and glided over the floor so gracefully that I kept thinking of the way a well-trimmed sailing ship moves across the water.

When I asked her for a dance she accepted. We had such a good time dancing that I wouldn't let go of her from then on, which didn't set very well with the fellow she had been dancing with earlier. He gave me some guff about it out on the dance floor. In those days I didn't believe in long arguments, so I just hauled off and knocked him down. That wasn't at all uncommon at those dances. Nobody made a fuss about it and he backed off from then on. We enjoyed all the dances for the rest of the evening, and I discovered that she was there with an old friend from the island, Edith Anderson. Having a friend in common like that made the evening even better. We spent a wonderful time together at the dance, then she went home to spend the night with Edith.

I was pretty hung over the next morning when a call came for me at my parents' home. My youngest sister, Ellen, came to tell me that a lady was on the phone for me. It turned out to be from the young lady at the party, whose name was Gwen. She told me what a nice time she'd had, and we wound up making a date to get together in downtown Seattle a few days later. We went to a movie, again enjoyed each other's company a great deal, and started dating regularly.

She was more beautiful than ever each time I saw her. I liked the clothes she wore, her manners, her sense of humor and the way she would laugh until tears came in her eyes. I couldn't get my mind off her. I finally had to admit to myself that I was falling in love.

The berry season ended about that time, when my father got a new contract to haul salmon from the fishing grounds to a cannery in Port Townsend, owned by a fishermen's co-op. Paul and I worked together with Dad on that job, taking me away from my new girlfriend, but we managed to get around that by having her along on some of the trips to the salmon banks. She had a flexible work schedule because she was employed by a Seattle author at the time, Frank Richardson Pierce, for whom she typed the manuscripts of his popular adventure books set in Alaska and the Old West. By the fall of 1930 the salmon season ended and I moved into the Stevens Hotel in Seattle with one of my friends, Ken Voight, in order to look for a job along the Seattle waterfront, meanwhile keeping steady company with Gwen. We had grown very fond of each other by this time.

It seems strange now to think that I could afford to live in a Seattle hotel while looking for work, but it actually was an inexpensive way to live then. The Stevens was located at First and Marion in Seattle, on the southwest corner of the present site of the new Federal Office Building. Ken and I shared a room in which we each had our own bed, and a quite decent room at that, splitting the cost of six dollars a week. The hotel even boasted a bellboy, who doubled as the local bootlegger. One day while we were staying there a federal agent approached the bellboy, asking for a bottle of liquor. When the bellboy produced it the agent identified himself, whereupon the bellboy hit the agent hard enough to knock him down and hightailed it out of there, never to be seen again.

This was the early part of the Great Depression. Times were tough on the waterfront. However, I finally found a job with Calmar, a subsidiary of Bethlehem Steel, on their ship the *Texmar*. She was a cruddy-looking ship with an East Coast crew that matched, but I needed the job and hired on as an AB. The focsle was locked when I went aboard, as they were afraid the crew would steal from each other if they could get in there. We went to work immediately lashing down cargo, working all night with no midnight lunch served. When we got under way for Tacoma the next morning they finally let us in the focsle. It was dirty, with blue denim sheets on the bunks instead of the clean white sheets we received weekly on the Griffiths ships. I was so disgusted with the ship that I quit when we got to Tacoma. The captain was furious, refusing to give me the two days' pay I had coming. Even though I needed the money I quit anyway, taking a bus back to Seattle.

Gwen was happy to see me, but I found things getting even worse as far as getting work. I continued living at the Stevens Hotel, looking for work, occasionally crewing for Dad on the *Hannah C*. This went on through the fall and early winter of 1931. Even the Griffiths Company, my old mainstay, was running into difficulties, with many of their ships laid up in the boneyard. They did haul one out once in awhile for one or two trips, using a crew of old waterfront hands. A couple of these seamen were to play important parts in activities of the Sailors Union of the Pacific. One was Harry Lundeberg, eventually to become union secretary; another was Albert Larson. They were good workmen and fine sailors. I got in with that gang, and in February 1931 I was called to serve with them as an AB on the *James Griffiths* when she was broken out of the boneyard for a couple of trips.

We hauled bulk grain down to Vallejo and then went in to the Howard Street dock in Oakland to load rock salt in bulk for Tacoma, delivering that to the Hooker Chemical Company. Fortunately for me, it was spring by the time the *James Griffiths* went back to the boneyard, and Dad was getting busy again.

The summer and fall of 1931 were a repeat of the previous year. Working the berries, following up after mid-July with hauling salmon for the Port Townsend cannery, making powder runs for Du Pont, we kept busier than ever. By now I was keeping company with Gwen as much as possible when not off somewhere with Dad on the *Hannah C*. I was fond enough of her to want to marry, but she wouldn't have that as long as I intended to follow a career at sea.

When that busy time ended I moved back to the Stevens Hotel again to look for work along the Seattle waterfront. This time, though, I had no luck at all. The Depression was really setting in. Even Dad was having trouble picking up the odd jobs he had done in earlier years to fill in during the winter. Once in awhile he'd have some freighting or towing job when he would need my help, but that was the only work I had. Times were tougher than ever. If I hadn't been a saving sort of person, with some money put away to get me through a period like that, I would have been dead broke.

As things turned out I didn't get any work that fall and winter other than a few trips crewing for my father. When March rolled around it was time to put the *Hannah C*. in dry dock. Dad took her back to the yard where she'd been built, Olson and Sundy in Ballard. With Paul also having fallen on hard times, he and I both went back to work for Dad, scrubbing and painting the boat's hull, and going over all the machinery. We took the heads off the engine to clean out all the water-cooling passages, removed the valve cages and took them to a machine shop to grind the valves, then lapped them into the valve seats ourselves. Finally, we painted all the topsides so that she was like a new vessel, painted from truck to keel.

Before putting the *Hannah C*. in dry dock we had hauled many loads of fertilizer to Bainbridge Island for the strawberry farmers to put on their fields. When the work on the boat was finished we did more hauling in preparation for the berry season, mostly sugar and barrels, and before we knew it the season was on us. It was even busier than the previous year, so that some nights we were hauling 2,000 crates of fresh berries and more than 500 barrels of sugared berries to cold storage. Dad was making good money again.

Paul and I had worked gratis for Dad when fixing up the boat, but he paid us well during the cargo work of the berry season. I got five dollars a day, a lot more than I could earn at sea as an AB. At the end of the berry season in 1932 Dad got a new job, working for a cannery in Southeast Alaska. When he offered me a job to go with him to Alaska I couldn't turn it down.

The contract was with the Diamond "K" Packing Company in Wrangell, located in southeast Alaska. Herb Kittelsby was superintendant of the company at the time; Carl Thiel, the owner. When they expressed an interest in chartering the *Hannah C.* we ran over to Pier 4 in Seattle so they could take a look at her. Both of them came down to the pier to examine the boat, which satisfied them completely. Dad had the contract.

Before leaving for Alaska about mid-July we loaded in Seattle for the trip north. Cases of empty cans and lids, general stores and a deckload of sacked coal for the cannery made up the load. The crew consisted of Dad as skipper, Paul as deckhand, myself as deckhand and engineer, another deckhand and a cook. With that crew and a couple of cannery people as passengers, we ran straight through to Wrangell in eighty hours. It's a beautiful run up there through the Inside Passage. Passing through Johnstone Straits, Queen Charlotte Sound, up through Fitzborough Sound, Frazier Reach and Grenville Channel, it was a trip I never tired of as I made it many times in later years. Dad, Paul, and I took turns on watch, Dad trusting Paul and me completely to navigate those passages.

After discharging supplies at the cannery, we started helping with placement of the company's fish traps. When Alaska gained statehood, one of the first acts of the original legislature in 1959 was to outlaw these traps, which had been a longtime irritation to fishermen. The inability of Alaskans to control their own fisheries while under territorial status was an important factor to voters when making the decision for statehood. But in 1932, while the fishery was still under control of the Federal Government, traps were a principal means of catching salmon.

There were two kinds. One was a floating trap. Diamond "K" had seven of these traps spread out through Southeast Alaska, each a square made up of floating logs with a small watchman's house at one corner. Wire webbing was hung from the logs to the bottom, and webbing extended from the trap to the shore, up to a quarter mile away, so that schools of salmon following the shoreline would meet and follow that lead webbing into the trap. The other kind of trap was a pile trap, similar to the floating kind except that piles were driven in place of the floating square of logs. Diamond "K" had two of these, one located at Kings Mill and one at Snow Pass.

It was a big job setting those traps out for the season, one requiring pile drivers, rigging scows and three tenders besides the *Hannah C*. A crew preceded our arrival to get the floating traps ready. We would tow them out to their sites, where the rigging crew would run a big wire shore line to the beach, making it fast to a strong tree or stump. Because of the strong tides in Southeast Alaska the traps needed up to six heavy anchors, each weighing about a ton and fastened to the traps by 1800 to 2000 feet of heavy wire cable. Part of our job was to tow the rigging scow out to drop these anchors. Even with all those precautions the anchors would sometimes drag during the season, when we'd again have to tow a rigging scow out to reset the anchors and tighten up on the anchor lines.

When all the traps had been set and the season opened, we went to work hauling salmon. Besides brailing salmon from the traps and taking them to the cannery, we'd pick up fish from gill netters and seiners. Some of those runs were pretty long. The company had three traps down in Clarence Straits, one at Narrow Point, one at Ratz Harbor and another at Olson's Cove. We brailed all those traps when we were down that way. Another long trip was to brail the trap at Kings Mill. Leaving Wrangell we'd go through Wrangell Narrows, past Petersburg into Fredericks Sound and down into Chatham Sound to reach the trap, a run of about ten hours that we'd usually make during the night. Sometimes if there weren't enough fish in the trap when we arrived, we'd lay over for a day in Security Cove until there was a good load.

I enjoyed that work with the cannery. We worked hard, often night and day, but there was always something different to do and Southeast Alaska is so beautiful in summer that the whole experience didn't seem like work.

When we laid up at night, tied to a trap or anchored in a cove, we'd often play a game of four-handed crib. We did this one night right after talking with the cannery on the ship-to-shore radio they had provided for our use on the *Hannah C*. The game had gone on for about an hour when somebody looked up and noticed that the transmitter light for the radio was showing "on." Whoever had used the radio had forgotten to shut it off before the game was started. Considering that Paul had a very loud voice, and used language not usually acceptable in good society, we were worried that the Federal Communications Commission would give us a citation. That never came, but when we arrived in Wrangell a couple of days later it seemed that the whole town was on the dock to meet us, asking when we'd have the next crib game. Someone in town had heard us on the radio when the game was started, had alerted just about everyone in town, and those who had radios tuned in to the program. We took a lot of good-natured kidding about that for a long time.

At the end of the season our work was just the reverse of that at the beginning. All the webbing was cut off the traps, discarded by simply letting it sink to the bottom. A rigging scow pulled the anchors on the floating traps, saving the cable by winding it on spools, and the traps were then towed in to safe harbor where they were pulled up as far as possible on the beach at high tide to be left there for the winter. All the pilings for the pile traps were pulled, and also put on the beach for the winter, as they lasted much longer that way than if left in place. The spooled anchor cable was stored at an old cannery at Lake Bay, where the spools were soaked in big vats of heated tar before being put in storage.

Considering all the work that had to be done it was an expensive operation. It cost $15,000 to $20,000 to drive the Kings Mill trap alone, a lot of money in those days. Some traps cost even more to put in, so it was a big investment the company made in a gamble that there would be plenty of salmon.

Our final job was to load as much canned salmon as possible on the *Hannah C.*, to take to Seattle on our return. Since we were going in that direction anyway, this saved the company freighting charges for that much salmon. As on the trip north, a couple of passengers from the cannery came along with us, saving the cost of travel on one of the passenger ships. We left Wrangell about the first week of October, running straight through again and arriving slightly more than three days later.

Once again I went back to live at the Stevens Hotel, and once again I had no luck at getting a job, filling in with occasional trips crewing for Dad. It was good in one way to be back, though, as I had missed Gwen while in Alaska and was very happy to be with her again.

While I was in Alaska Gwen had taken a job as credit manager for the Ben Tipp jewelry store, giving her a steady income. With my savings from working for Dad I got along pretty well despite being out of work. We did quite a bit of partying in those days with friends such as Nelda and Ed ("Johnny") Walker, who lived out in West Seattle. Partying didn't cost a lot in Depression days. We usually had enough money to take a bottle or two of happy juice out to Nelda and Johnny's place when we visited. They were a nice couple and we had something in common in that Johnny had served many years on President Line ships as second mate, now working as supercargo at the Matson Line office in Seattle.

Gwen and I each had a car at the time, she a Nash roadster while I had a 1928 Chevrolet that I had bought new. Since times were hard and we were with each other all the time anyway, we decided that I would sell my car and bank the money. I sold it privately for $150, not much considering that it was in nice shape. While getting ready to sell it I had become acquainted with a fellow in the used car business, Johnny McGee, who had answered my ad for the car. I didn't sell it to him because he was a very sharp dealer who had to be watched like a hawk, but we did become friendly.

About the same time, Gwen bought a couple of raffle tickets from a girl who worked with her at Ben Tipp, paying 10 cents each for the tickets on a raffle sponsored by one of the local synagogues. She had forgotten all about the tickets when one day Mary Neff, the girl who had sold them to her, came into her office all excited. "You won it, you won it!" Mary told her. "Won what?" Gwen asked. "You won the refrigerator on the raffle!"

Gwen didn't know what to do with the refrigerator. She had recently moved to a unit in the Lexington Apartments in downtown Seattle which was furnished with a refrigerator. When she asked me to pick up the prize for her, I called a teamster friend of mine, Joe Francis. Together we went up to the synagogue to get the refrigerator and move it into Gwen's apartment. Some of the businessmen at the synagogue offered to buy it, but Gwen and I decided that she might do better. We advertised it in the newspaper in trade for a car.

Lo and behold, it was Johnny McGee who called to say that he had a car he would trade for the refrigerator. He offered a nice Chevrolet coach in the trade, which we accepted. Being the wheeler and dealer that he was, when Johnny learned we didn't really need another car he offered to sell it for us. We set a price on it and he sold it for that price, giving us an interest in the used car business. Talking to Johnny, we learned that many of the local dealers were going back to Chicago to buy cars and bring them out in caravan for resale in the northwest. So Gwen and I decided we'd pool all our money, $800 altogether, and that I would take that money back to Chicago to try my luck at buying used cars.

I paid $22 for a bus ticket to Chicago. That and money for food along the way came out of the $800, as did the cost for a hotel in Chicago. After scouting around Chicago for a couple of days I found that I could buy three cars and still have enough left over for expenses to drive them back to Seattle. I bought a tow bar to pull one car behind the other, but needed a driver for the third car. A lot of kids hung around the lots in those days, hoping for a job to drive out west. It wasn't hard to find a couple of them to go along with me, driving one of the cars. We set out on the trip west very early in the morning, driving until late each day, when we'd stay in an auto court and start early again the next day. It took a little less than a week to get to Seattle.

We put the cars on Johnny McGee's lot in Rainier Valley, offering them for sale with credit terms because of Gwen's experience in managing credit. Johnny got all three of them sold, and with the twenty-five percent annual interest rate collected we ended up just about doubling our money on them. It took a while before we got all that money, though, and in the meantime I really needed a job.

At Thanksgiving, my Mother wanted me to bring Gwen over to the island to get to know the family. She and Gwen hit it off right away, having such a good time together that Gwen was invited back with me for parties at Christmas and during the New Year holiday. With family and friends around, and Gwen beside me, I remember those times very fondly.

Shortly after that, in early 1933, Gwen and I visited the Walkers' home in West Seattle. While we were there Johnny mentioned that one of the Matson ships, the *Maunalei*, was dry docked in Seattle. He had heard a rumor that a job might be open on the ship. Johnny called not long afterwards to tell me that there was an AB job open on the *Maunalei*. He said he'd already talked to the mate about me, and if I got right down to the Todd Shipyard I could have the job. I hustled down there to report to the chief mate, Mr. Davis, who took a look at me and said I had the job. I was happy to be going out to sea again.

She was a big steamship, carrying eight AB's split evenly between Hawaiians and non-Hawaiians. I got friendly right away with a fellow named Duke York, who turned out to be the only other member of the crew belonging to the Sailors Union. We tried to talk the rest of the deck crew into joining the union, but they wouldn't have anything to do with it.

The *Maunalei* was still in dry dock when I reported aboard. We worked on the rigging for the few days she was still in the yard, then loaded cargo in Tacoma and Seattle. Our last stop before leaving port was off Four Mile Rock near Seattle, where we were to load blasting powder. It was quite a feeling to see the *Hannah C.* coming towards us with the cargo, Dad at the helm. After the powder was loaded and I'd said goodbye to Dad, we set out for Honolulu.

There were two watches on this ship, six hours on and six hours off, with the Hawaiians standing one watch and the rest of us the other. When we arrived at the islands the entire deck crew went on day watch together, and we got a little time to visit ashore in the evenings. Although we didn't work cargo on this ship, we were constantly kept busy and often worked the clock around. The *Maunalei* was a triple-deck ship, having a full 'tweendecks and an orlop deck. With five hatches on the main deck, and five on each of the other two decks, it seemed we were constantly removing and replacing hatch covers while the cargo was being loaded and offloaded.

They really worked the crew hard on those Matson ships. There was no overtime even though we might work as long as twenty-four hours straight. Sailors were cheap then, earning $60 a month with no way of getting more through overtime pay. After leaving port we'd keep working together as one crew, securing gear and generally getting ready for sea, then go back to the six-and-six watches with a split crew. They had a gimmick in those watches which supposedly was to give the sailors a break, an arrangement that every third night a sailor didn't have to stand a watch on deck. It wasn't much of a break, though, because instead of working on deck we worked that "night off" scrubbing and painting in the quarters and alleyways below deck.

Matson was the only company I served that had time clocks aboard the ships. On the *Maunalei*, with its four-man watch at sea, one man stood wheel watch, one served as lookout, and another was "taking it easy" working below deck. The fourth man patrolled the ship, punching time clocks in various locations and having to turn the ticket in to the mate at the end of the watch.

Once again there was no midnight lunch provided for the sailors on watch, not even coffee. At least the galley was open, even if everything was locked up, so I would buy a can of chocolate before leaving port and go in there to fix myself a cup of hot chocolate for the night watch. It was a long wait until breakfast, and then that wasn't much. The *Maunalei* was a poor feeder. Breakfast was limited to mush, with a few cans of condensed milk diluted down until it was awfully thin. We did get a break from the ship's food once, while the galley was being refurbished during a layover in San Francisco. The sailors were given compensation money for eating ashore, not much, but it was at least a change.

One time during that first trip, as we were in good weather between Hawaii and the mainland, they put both crews on a 6:00 AM to noon watch to paint the lifeboats. Duke York and I, who were working together, decided about 10:00 AM that we needed a cup of coffee. We put down our paint brushes and headed down towards the galley. "Where are you guys going?" the mate asked. "We're going down for a cup of coffee," we replied, still walking in the same direction. "All right," Mr. Davis said. With that the rest of the crew put down their brushes and started to follow us. When Mr. Davis asked them where they were going they replied that they were also going for coffee. "You fellows don't deserve coffee," he told them. After that Duke and I regularly took a coffee break, while the rest of the deck crew was never permitted to do so. Talking it over, Duke and I decided that Mr. Davis let us do this because he knew we were union men and was himself sympathetic to the union. I got a chance to talk with him sometimes while we were on watch together, and while he never said so directly our talks also led me to believe he had a sympathetic feeling for the union's causes.

The *Maunalei* was on a "triangle run," going from Seattle to the Hawaiian Islands, then to San Francisco and finally back to Seattle. Leaving Seattle we always had a full deckload of lumber, and on top of that stalls full of cattle. A couple of cowboys came along to take care of the cattle and feed them. One time while it was still cold in early winter, we had hogs in the stalls instead of cattle. I remember how those poor hogs suffered in the cold, squealing all the time. In the Hawaiian Islands we always called at Honolulu first, then went on to other island ports such as Hilo, Port Allen, Kahului, and others, discharging general cargo and loading sugar. The ship returned to Honolulu after calling at those other island ports before starting out across the Pacific Ocean for San Francisco.

It was in March 1933, after we had left Honolulu bound for San Francisco, that while sailing in good weather we hit a series of very high, steep waves. They weren't cresting, but they were so large and steep that they broke over the bow of the *Maunalei*. We learned later that they were caused by a serious earthquake that hit Long Beach, California, on March 10th of that year, killing more than 100 people.

The *Maunalei* had side ports for loading, which were closed and bolted with gaskets when loading was finished. Mr. Davis was worried that some of them might have leaked when we were hit by those waves, so he had me go below with him to see if any water was coming in. The holds we were checking had been loaded with sacked sugar, piled full except for some crawl spaces left just for the purpose of this kind of inspection. With Mr. Davis ahead of me in one of the crawl spaces my only view was of his large stern end. Then I saw as he was crawling along that he had a big six-shooter strapped to his belt. We finished the inspection, seeing that there had been no leakage. I never did mention to him or any of the crew that he was carrying a side arm. It was the first time in my career that I had seen anyone armed.

Much of the cargo from the islands was discharged in San Francisco. Even though we kept up our regular day watch in port, one man was always assigned to a night watch, punching the time clock as he went around the ship. On the way up the coast anyone who wasn't standing a wheel watch or acting as lookout had to go down into the holds to clean them out, sweeping and piling up dunnage. Arriving in Puget Sound everyone was called out way ahead of the time we'd dock, getting lines out and ready, taking off tarpaulins and hatch covers, no matter whether or not we'd just finished a six-hour watch. We worked and worked on that ship.

These were the kind of conditions that were leading up to the seamen's strike the following year. But it was a job, with even that $60 a month looking pretty good in the midst of the Depression.

After making several trips on the *Maunalei*, I got off in early May in order to go back to work with my father. Even though we had been worked hard, I had found it an interesting experience and had no complaints about the officers. I had grown to like the chief mate, Mr. Davis, so I was sorry to learn later that he had lost his mind not long after I left. When the *Maunalei* called at Port Allen the bow was made fast to some large mooring buoys, the stern of the ship then swung in towards the beach. While the crew was going through this process, Mr. Davis went crazy and jumped overboard. Being a stout man he floated, so the crew was able to rescue him. I never did hear what happened to him after that, but I've often wondered if he hadn't cracked under the pressure of so much work, and having to drive his crew so hard for so long.

Duke York left the ship with me in Seattle. We shook hands, said goodbye, and never have seen each other since. I enjoyed Duke's company on the *Maunalei*, and have wondered from time to time what ever happened to him.

Paul had also been sailing while I was on the *Maunalei*, quitting his ship as I had in early May in order to work for Dad. The routine of working with our father during the summer was a real opportunity for us through those Depression years. Not only did we receive better pay than we could get otherwise, but we enjoyed being with Dad on all those usual activities of the berry season. And Dad enjoyed having his boys along with him. The Diamond "K" Company had been happy with our work the year before, renewing the contract with Dad to work out of their Wrangell cannery again this summer. We left for Alaska shortly after the Fourth of July holiday, doing the same kind of work as the previous year and returning in October.

Following the routine of earlier years I moved back into the Stevens Hotel with Ken Voight, kept company with Gwen, and looked for a job on a deepwater ship. The fall of 1933 saw conditions even worse than before as far as locating work, so I was quite discouraged until I finally got a small break in early November. Captain Hans Anderson, an old sailing vessel master and a friend from Eagle Harbor, came to visit me at the hotel one day.

"Poker Hans," as he was called, had served many years as master of the *Eleanor H.*, a four-masted top'sl schooner. When she was laid up along with most of the sailing vessels, he and his family lived aboard her at Eagle Harbor. He had then gone as skipper on the *John A.*, a little three-masted schooner owned by the Shields Codfish Company, sailing her to the Bering Sea for codfish, a rough fishery done entirely out of unpowered dories.

When Captain Anderson came calling on me at the hotel it was with the offer of a temporary job. He was going over to Poulsbo to take the *Phillis S.* out to Cape Flattery, to pick up the schooner *Charles R. Wilson*. They needed an engineer, and since the *Phillis S.* had an Atlas Imperial diesel like the *Hannah C.* he wondered if I'd like the job. I was happy to take it.

The *Charles R. Wilson* was in the cod fishery. These schooners would bring a full load of cod from the Bering Sea down to Puget Sound at the end of the season, discharging the fish at Poulsbo, then go to San Francisco for a cargo of salt to be used during the following season. She had picked up her load of salt and was on her way up the coast, due off Cape Flattery in the next day or so. When we got to Poulsbo I found the engine on the *Phillis S.* in quite bad shape. It took me about three hours to pull and clean the injectors and re-time the engine. When I rolled it over after that it ran fine. Hans was happy that he'd found a good engineer.

On the way up to Cape Flattery it was foggy all the way, but we found the *Charles R. Wilson* all right and towed her back to Seattle with no problem, taking her in through the locks to Lake Union. Besides Hans and myself there were two deckhands I knew well, old shipmates of mine. One was Heine Lang, another a big Finn named Joe. Just as Paul and I sailed deep sea in the winter and with our Dad during the summer and fall, these fellows spent the summer and fall working for Shield's company in the cod fishery and then went to sea during the winter. After we had brought the schooner into Lake Union, Heine and Joe were laid off. Mr. Shields asked Hans how I had done as an engineer. Hans told him I was good enough that he'd better keep me on until the arrival of a second schooner, due in a couple of weeks from San Francisco. I was happy about that. Earning five dollars a day while working in the heart of Seattle and living at the Stevens Hotel was right up my alley. I spent the time working away on the boat, cleaning up the engine and generally getting things into nice shape.

The second schooner was the *Sophie Christensen*, a four-masted schooner also loaded with salt from San Francisco. Heine and Joe were hired back as the deckhands when she was due to arrive. The weather was miserable while we again headed out to Cape Flattery, foggy and with a heavy easterly wind blowing down the straits. Approaching Neah Bay about midnight in those conditions, with visibility just about zero, we piled up on the rocks of a reef off Wada Island. There we were, bouncing from rock to rock in a sea of kelp. Heine Lang opened the door to the engine room and yelled down to me: "We're done for! We're done for! We can't get out of this mess!" I thought we for sure were goners.

The skipper kept me busy on the engine, backing and filling, trying somehow to get out of there. Frankly I don't know how, but somehow we did get out. The Good Lord must have been with us. Instead of heading out to Cape Flattery we found our way into Neah Bay, running slowly because the propeller was full of kelp picked up while we were on the reef. It was a terrible night! The heavy seas, wind, lack of visibility, and experience grounding on the reef made it a nightmare. Tying up alongside a barge at Neah Bay we waited until daylight to survey the damage. Amazingly the only damage was that the keel was chewed up a bit from the rocks, and the propeller loaded with that kelp. There was a kelp-cutting pole aboard the *Phillis S.*, as she was actually a big seiner used by Shields in the Alaska fishery, so we spent several hours clearing away the kelp. After that I went down and started the engine, finding that everything still ran fine.

We then headed out to the cape, the easterly wind still blowing strong, where we found the *Sophie Christensen* sailing along in fine shape. With that wind blowing so hard out of the straits it would be difficult for the schooner to head up into it. We thought we could help out by putting a line on her to keep her pointed higher into the wind than she could do by herself. The plan was to sail on a tack towards Vancouver Island and then, when close to the island, we would help her come about on the other tack. When that time came we took a strain on the hawser to bring her around. She started to come around, got caught in irons, then fell off on her original tack.

This caught us perpendicular to her course. With the towing bit almost amidships on the *Phillis S.*, and the tremendous power of those big sails on the schooner, we were suddenly being pulled sideways out of control, one rail underwater and the whole boat about to be pulled over. Joe, the big Finnish deckhand, grabbed an ax that was handy and with one swipe cut that heavy hawser. The *Phillis S.* popped up like a cork. We were saved, but there was the *Sophie Christensen* sailing off towards Vancouver Island.

Back we went to Neah Bay to assess our situation. There didn't seem to be much damage. Captain Anderson said he knew what the schooner would do. He figured they would make for the lee of Cape Flattery, probably anchoring in Mukkaw Bay, so we headed out around the Cape and found that was just what they did. Tying up alongside the *Sophie Christensen*, we waited about a day until the wind calmed down and then towed her back to Seattle. That had been quite a trip, with two incidents where we might have lost our lives.

With no more schooners to tow, Mr. Shields laid us all off after a couple of days back in Lake Union. It was so close to the holidays then that I didn't look for work right away, enjoying the celebrations over Christmas and New Year and then starting to look for work in early January of 1934. There just wasn't any job to be had. By then, Gwen and I had pulled together enough money from our work and the payments on the cars sold earlier that we had a stake for another trip to Chicago to buy more cars. It was a pretty miserable trip, with snow and ice all over the Midwest, but I managed to buy four cars and get them out to Seattle with the help of another pair of kids wanting to go west. Again we put them on Johnny McGee's car lot and sold them on time payments.

It was sometime in February by the time I got back to Seattle with the cars. I spent the rest of the winter looking for work, hanging around Johnny McGee's lot as the cars were sold, frustrated by the Depression and lack of work. It was another year with no deep-sea experience. When spring came I felt lucky to go back to work with Dad, getting ready for the berry season.

In May, we had the *Hannah C.* in at the Pacific Steamship dock at the foot of Connecticut Street in Seattle, ready to take 2000 sacks of sugar off one of their freighters. Dad went up to the office to make the arrangements, coming back to tell us that the longshoremen were on strike. An official in the office told him the strike would probably be settled in a couple of hours. We laid there all day, waiting for the sugar. Finally one of the officials came down to tell us to get the sugar from a supply already offloaded onto the dock. It looked like the strike would not be settled soon.

⚓

CHAPTER TEN

The Strike and a Far East Experience

THE *PRESIDENT JEFFERSON*

"You can put me in jail, but you cannot give me narrower quarters than as a seaman I have always had. You cannot give me coarser food than I have always eaten. You cannot make me lonelier than I have always been." — Andrew Furuseth

Sea Travels

THE AUTHOR of these touching lines, holder of Card No. 11 of the Sailor's Union of the Pacific, was elected secretary of the union in 1887 and served in that position for the next fifty years. My father, who joined the union soon after arriving on the West Coast in the early 1890's, knew and admired Furuseth, and he certainly knew the sailor's hardships that led to the writing of these lines.

While I neither knew Furuseth nor quite the harsh conditions Dad had experienced, the conditions for sailors in the 1920's and early 1930's were bad enough for me to sympathize completely with the sailors' strike that was called immediately after the longshoremen walked off the job on May 9, 1934. Contrary to what dock officials thought when the strikes were called, they were not settled quickly. It was not until eighty-three days later that sailors and longshoremen went back to work. All shipping was tied up for those almost three months, including the vessel from which we had planned to take sugar.

Once again I was lucky to be working for my father, as the strike didn't affect local hauling around Puget Sound. We felt awfully sorry for those out on strike who manned the picket lines. There was no strike pay for them in those days. They were entirely on their own, rustling up food however they could, setting up makeshift kitchens near the picket lines to cook whatever food was rounded up. Dad, Paul, and I contributed to them when we could. When we ran into Harry Lundeberg we'd have him send somebody down to the dock to pick up a couple of crates of strawberries, which we'd buy from the farmers on Bainbridge to donate to the strikers.

One time during the summer, we got involved in a militant action. Harry Lundeberg called at my folks' place on Eagle Harbor just at the end of the berry season while we were getting the *Hannah C.* ready to go north, asking if we could get a gang together. A bunch of strike-breakers were headed over on the Port Blakely ferry to take out two ships laid up at the old mill dock. "Try to stop that gang of scabs before they get off the ferry!" he said. The fellows on the picket line had built up a good intelligence system, so they knew whenever strike-breakers were going to try to take out a ship.

Paul and I, and our friend, Yankee McLaughlin, rounded up some other friends and got over to the ferry dock at Port Blakely. Carrying baseball bats and clubs, we stood at the head of the dock as the ferry arrived. The gang of scabs started up the dock as they got off the ferry, took one look at us and turned around to get back on. They never did take those ships out.

It was nothing but hard times for those out on strike and without work that summer. In the midst of the Depression, with no strike pay, no unemployment insurance, no welfare, lots of them just plain went hungry.

Dad, though, had his usual contract with the Diamond "K" cannery in Wrangell, so we were fortunate to be along with him again in that summer of

The Strike and a Far East Experience

1934. After loading general stores for the cannery and taking on a deck load of coal in Lake Union, we picked up a small barge to tow north. The season in Alaska was much like all the others. By the time we returned about the first of October both the sailors' and longshoremen's strikes had ended.

Gwen was due for vacation at that time. We decided to combine her vacation with business by going back east to buy more used cars, this time together. Somebody had suggested that we go to Detroit instead of Chicago on this trip, which proved to be a bad idea. We could have done better buying cars in Chicago, but we bought four in Detroit anyway. I hooked two of them together, driving that pair myself while Gwen drove one and we hired a westward-bound kid to drive the last.

By the time we got through Chicago on the way west, Gwen was getting tired of driving. I kept my eye out for hitchhikers, looking for one who might be suitable to drive that car instead of Gwen. Seeing a likely prospect, I stopped to pick him up. "Oh boy," he said, "I'm glad to get a ride. How far are you going?" "To the end of the line," I answered. When he asked what I meant by that I told him we were headed for Seattle. "Whoopee!" he shouted, taking off his cap and throwing it down on the floorboard of the car. "I'm headed for Everett. This really is my lucky day!" He explained that he had a job in Everett if he could find a way to get there, and was ready to ride the rails if he couldn't find any other way to Everett.

The hitchhiker turned out to be a good find. He was a nice fellow and could drive, so after we stopped for lunch Gwen turned her car over to him and we were able to make the rest of the trip together. We bought meals for the kids driving for us and all stayed in auto courts along the way after driving long hours. They were such nice kids that I slipped each of them five dollars at the end of the trip, which they appreciated very much. In those hard times it was a break for them to travel that way, and a break for us to find nice people to help drive our cars out to Seattle.

By that time it was November 1934, and I was anxious to get back to sea. One of the changes brought about by settlement of the seamen's strike was that now hiring was done out of the union halls. I went down to the Sailors Union, registering and getting a card with a number on it. Later that month I got a call from the union dispatcher, learning that there were three AB jobs open on the *President Jefferson*. I threw my card in for one of those jobs, got it, and in early December went down to Pier 91 at Smith's Cove where the President Line ships docked, reporting to the chief mate, Mr. Stull. He was a fine fellow, formerly a skipper of a number of smaller ships in the Alaska Transportation Company fleet.

The *President Jefferson* was a big, first-class passenger and cargo ship, with a crew of more than 200. When the jobs were passed out after I reported for duty I was appointed "station man", a good job. In that position, I was responsible for keeping the bridge in good order, relieving the quartermaster at the wheel when he needed a break during the watch we took together, and being lookout while running at night. Sometimes I'd stand a gangway watch for passengers boarding or disembarking by one of the two decks where they had access.

Another of the station man's jobs was to take care of all the flags a ship carries. I raised the ensign at the staff aft in the mornings while in port, taking it down again at night. On leaving port the ensign becomes the steaming flag and is raised to the gaff on the mainmast. We carried the Naval Reserve flag and the Mail Flag at the yards, and I also took care of all signal flags.

During the early morning watch from 4:00 to 8:00 AM, the deck crew scrubbed and cleaned the nice teak decks and rails in the passenger area, making them ready for the 500 passengers who would soon be starting their day at sea.

Our first port after leaving Seattle was Yokohama, reached after quite a rough voyage during about eight days across the Pacific. The *President Jefferson* was a fast ship, able to steam at about twenty-four knots, though on a long-distance cruise like this she probably sailed at closer to twenty knots. Following a great circle route we were able to see some of the Aleutian Islands before the Japanese Islands came in sight.

After discharging some passengers and boarding others at Yokohama, and handling freight, we went on to do the same at the Japanese ports of Kobe and Nagoya. The ship's next port of call came after a voyage through the Inland Sea and into the mouth of the Yangtze River, when we reached Shanghai. At the Japanese ports both passengers and cargo were discharged and loaded at docks, but at Shanghai only the freight was unloaded at a dock. The ship was anchored in the middle of the stream during most of the time in port, a barge tied alongside as a platform for the gangway and a landing for the launches used to take passengers back and forth to shore.

Another of my jobs as station man was to see that everything aboard was locked up properly while we were in port, such as the paint lockers and the bosun's locker. One morning while in port at Shanghai, the chief mate called me up and asked "Christensen, did you lock up the storeroom last night?" "I sure did, Mr. Stull," I answered. "Well," he said, "they cleaned it out last night and the padlock is gone." When I went down to check it out I found they had actually taken every last thing out of the storeroom. I felt bad about it because I was always sincere and careful about my job, but the Chinese were very clever about getting things off a ship. Laying in Shanghai we had to keep the portholes closed because otherwise thieves would come along in sampans and poke in through open portholes, using long poles with hooks on them to steal clothes

The Strike and a Far East Experience

and anything else they could snag. While at the dock there to work cargo, we had to keep fire hoses on deck on the offshore side of the ship to drive off sampans trying to come alongside to pilfer whatever they could off the ship.

In Shanghai I found an old shipmate from my days sailing on Griffiths Company ships, an Englishman named George Folks. George sailed with the British Navy in his earlier days, had deserted in Alexandria and become a merchant marine sailor, wanted by British authorities. His influential father had somehow arranged a King's pardon for George, who continued to sail until reaching the Chinese coast. He left his ship to join the Shanghai River Police, becoming a sergeant by the time I met him there. George spoke Chinese well, had a car, belonged to a club, and knew the city as well as anyone, so it was a real pleasure for me to be escorted around Shanghai by this old friend. I also got some lessons in what the world was like outside my own country.

On our outings George would come out to pick me up from the ship in a river patrol boat, making me feel pretty important. One of the lessons I learned came during the first time I visited Shanghai with George: that the hard-working Chinese longshoremen were slaves in their own country.

As George and I were leaving the ship to walk across the barge to his launch, the longshoremen were unloading hatch covers and other canvas from the *President Jefferson*, throwing them from deck down to the barge. One of the Chinese on the barge had picked up a tarp to move it, and accidentally hit George in the leg as we walked by. George's response was to kick him hard in the butt, giving him a string of Chinese along with it. The poor Chinese longshoreman was cowering when we walked on to get in the launch. "What did you do that for?" I asked George as we got in the boat. "Oh, you have to do that to get their respect," he answered. "That's a hell of a note, to have to respect some foreigner in your own country," I said. Being good friends we didn't argue about it, though George didn't think much of my attitude.

The first thing we'd do ashore was to go up to his club, a place where the river police entertained and gathered to visit with each other. After having a couple of their standard drinks there, scotch and soda, we would go out in his car to see the city. George was a terrible driver. On one trip through the city he hit a rickshaw, knocking the poor rickshaw man up onto the sidewalk. George stopped long enough to cuss him out before we continued our sightseeing tour. "You're treating these people like hell around here," I told him, "someday they're going to rise up against you." "He shouldn't have been there," was all George said.

I had never known George to be a mean man when we had been shipmates. It seemed almost that he had been trained to be this way as a British colonialist. Other things about the situation in China bothered me during my visits there, one of them being the Japanese occupation of some of the districts around the International Settlement in Shanghai. They had stormed Shanghai with 70,000 troops a couple of years earlier, establishing a "demilitarized zone" around the settlement. We would see Japanese troops patrolling around those areas on motorcycles with small guns mounted on them. Sometimes the Japanese would fire at junks in the river, and occasionally on ships of nations they considered unfriendly, though they never bothered us. It appeared to me that the Chinese had no say at all in their own country.

Another pathetic thing was the number of beggars around the Shanghai Bund, the embankment in the International Settlement that was a popular gathering place. Most of them were little children who would try to shake down the foreigners for money or anything else they could get. "No papa, no mama, poor little son-of-a-bitch," they would say in the English they'd picked up on the streets. Having heard that story, when a little boy approached me one day with his hand out I asked "What's the matter, no mama, no papa?" Looking sad, he rubbed his belly and said "No chow."

There were a lot of people with no chow in the Shanghai of that time. Garbage from the ships was dropped overboard right into the river, something that seems shocking now with so much concern about the environment. From a passenger ship like the *President Jefferson*, there was a lot of good food going to waste. When it was time to dump garbage, junks and sampans swarmed around, using dip nets to fish it out of the river and sometimes fighting over the better stuff.

Except for the ship's own cranes, longshoring was done strictly with labor. Moving heavy timbers, for instance, was done with cant hooks. Several people, each with a cant hook, would work in pairs to move the timbers with brute force. The longshore crews included women as well as men. Those Chinese women knew all about "equal rights," even back then.

Altogether the treatment of the Chinese, the Japanese occupation and the extreme poverty had quite an effect on me. I came to appreciate the old saying, "he doesn't have a Chinaman's chance," and to understand why Mao Tse Dung was able to gather such a following among the people.

From Shanghai the *President Jefferson* sailed to Hong Kong, adding to the crew for this part of the voyage a number of Chinese who did painting and other maintenance work. A second maintenance crew came aboard at Kowloon, Chinese who specialized in fabric work. They patched and sewed lifeboat covers, ventilator covers, awnings and the like. These work crews were necessary because our own crew had all we could do to keep the ship going. She was a big ship with lots of gear aboard, including thirty-two cargo booms. Coastwise people would call her a thirty-two-boom steam schooner, too much of a ship to keep in top condition just with the operational crew.

During our stay in Hong Kong's harbor we had a chance to visit Kowloon, where the ship docked, and sometimes took the ferry over to Hong Kong to see that side of the bay. During one of the visits to Kowloon I bought a beautifully carved camphor-wood chest, a lovely piece that I still own. The chest cost ten dollars. I had to pay an additional fifteen cents to have it transported down to the ship. On getting back to Seattle, it cost half the price of the chest — five dollars — to have it moved from the ship up to my apartment.

The last outbound port of call for the *President Jefferson* was Manila, the turnaround point for the voyage. A couple of days were spent discharging and loading cargo and passengers there. Then we retraced our course to call at Hong Kong, Shanghai, and the Japanese ports before returning to the Pacific Northwest with a call at Victoria before arriving in Seattle. We were fed well on that ship. Coming off a watch at midnight we'd go to the galley to get what they called a "black pan", with all the fancy cakes and cookies that the passengers enjoyed. There was a large crew of stewards, eighty-five Chinese who acted as waiters and room stewards, and had quarters of their own.

The master of the *President Jefferson* was Captain Lusty, a very stern man who ran a good, tight ship. He was a classic passenger ship master who knew how to take care of his passengers and entertain them. One morning I was on watch on the bridge when he came up to take a look around. Someone had left a baggage truck on deck. "Christy," he said to me, "you go down and give Mr. Stull my compliments, and tell him we're running a goddamn passenger ship. I want that freight truck off the deck." I paid the skipper's compliments to Mr. Stull, who saw to it that the truck was moved immediately.

One of the captain's attempts at keeping the passengers happy really fizzled. A couple of days ahead of a scheduled lifeboat drill he announced in the ship's newspaper that during the drill there would be a practice firing of the Lyle gun, used in emergencies to send a line across a great distance to another ship. He stirred up such interest that most of the passengers were out on deck to witness the event. The crew loaded the gun with gunpowder, stuffed some wads of paper on top of that and then put in a dummy projectile. The moment came, the fuse was lit. There was a faint little puff from the gun, the projectile came slowly out of the gun and barely cleared the rail before falling limply into the sea. Sometimes, as I discovered later in my career, there are things that are simply beyond the control of a captain.

Another incident beyond Captain Lusty's control took place at Kobe. There wasn't much in the way of sewers in the city, so they had a system of carrying the excrement out to sea in barges pulled by tugs, where a false bottom was released and the stuff was dumped into the water. There was always one crewman on the "honey barge," as they called them, to handle lines and the dumping. It's not hard to imagine what kind of job that was, standing alongside all that human dung.

One warm afternoon we were heading in through the breakwater at Kobe with a Japanese pilot on the bridge, when one of those honey barges got in our way. Having no room to maneuver in those tight quarters, the *President Jefferson* cut the barge exactly in half, leaving the pipe-smoking Japanese crewman standing on one half still floating, shaking his fist at us. This happened just as the passengers were sitting down for lunch. With the portholes left open because of the warm weather, the awful stench from the "honey" floated up through the dining salon. All those passengers rose as one person, getting out of there as quickly as they could to find some fresh air.

If a captain can't control everything that goes on aboard his ship, he certainly can't control the weather. We hit a typhoon in the South China Sea that raised hell aboard. We almost lost all the lifeboats on one side of the ship, managing to save them only through lashing them down tightly but losing the lifeboats covers and other canvas. A storm like that doesn't do much for passenger comfort.

A captain isn't always right, though. Sometimes even Captain Lusty was embarrassed by something that should have been under his control.

The *President Jefferson* carried a cadet in the crew, a son of one of the owners who was learning the seagoing trade. He spent time on the bridge as an observer, and sometimes worked with the deck crew to learn that work. The cadet happened to be on the bridge one day as we were docking in Yokohama, with a stiff wind blowing offshore. The wind was blowing so hard that an anchor had been dropped on the offshore side of the ship, with tugs trying to push her into the dock. She wouldn't budge, still laying there a long way out. Captain Lusty was stomping around on the bridge, wondering why his ship wouldn't move even with all the power of those tugs. "Excuse me, Captain," the cadet said, "but maybe if you'd slack off on the anchor chain she'd move into the dock." Sure enough, when the captain looked forward he could see that the anchor chain was taut, with the tugs trying to push against the almost unmovable anchor gear.

That young man had his own share of embarrassment, also while the ship was docking. Working with the deck crew that day, he was responsible for getting a messenger line ashore to use for dragging the big hawsers to the dock when we were close enough. The messenger line had been hanging along the side, and when the cadet pulled on it he found it had somehow snagged near one of the stateroom portholes. Going below, he located the stateroom and knocked on the door. Two girl passengers opened the door, dragged him in and locked it. They had seen the line by the open porthole, brought it in and taken a turn around one of the porthole fastening dogs. It was some time before the cadet managed to get the messenger line loose and himself out of the stateroom, and he lived with a lot of kidding after that about being the only crew member to be raped aboard ship.

Looking back on it now, I realize that my time on the *President Jefferson* was remarkable in many ways. I got some valuable experience on one of the best large passenger ships of the time, and had some fun along the way. But I also had a chance to see the Far East as the colonialism it had known for so long was going through its last days. The treatment of the Chinese in Shanghai, and the presence of the Japanese soldiers there were memories that stayed with me during the very few years left before World War II, and had a special bearing on the next seagoing job I would take.

⚓

Sea Travels

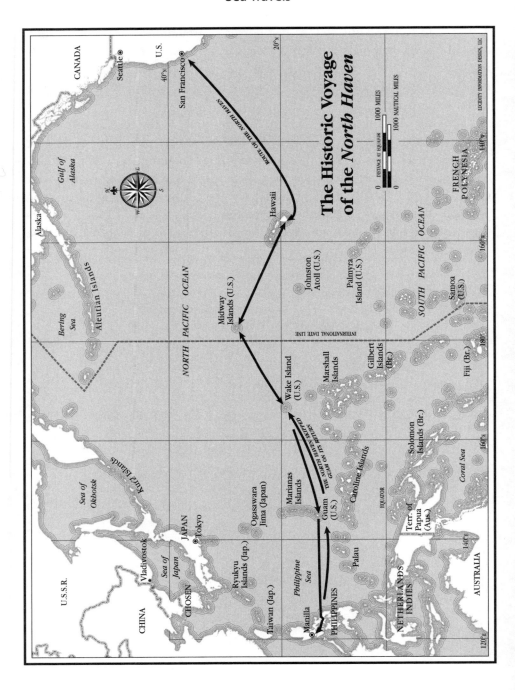

THE VOYAGE OF THE *NORTH HAVEN*

CHAPTER ELEVEN
An Historic Voyage

THE *NORTH HAVEN*

IT WAS MARCH 1935 when the *President Jefferson* returned to Seattle at the end of my second trip. As that trip ended, I found myself anxious to spend more time with the love of my life. Gwen and I had too little time together during the short turnaround time in Seattle.

My father solved the problem again with the offer of a job on the *Hannah C.* Not only was the pay better than on the *President Jefferson*, but this year Gwen used her vacation time to spend a couple of weeks with us in Alaska.

She traveled to Alaska on the *Northland* with my old friend Ole Monsass sailing as chief mate. A mutual friend, Bill Henshaw, took Gwen down to the boat when she left Seattle, introducing her to Ole so she'd have somebody to talk with on the trip north. Ole soon discovered that she had packed more for a cruise on a big ocean liner than for spending time on a cannery tender, with all kinds of dresses and "six straw hats", as he exaggerated in telling the story later. When she left the ship at Wrangell, Ole talked her into leaving the dresses and hats in his quarters on the *Northland*, taking slacks and more practical clothes for her time on the *Hannah C*. Her clothes remained there until Gwen boarded again for the return trip to Seattle a couple of weeks later, bringing Ole a lot of teasing for having lady's clothing in his locker.

The weather happened to be good during those weeks of August Gwen spent with us, making the time even more enjoyable than we had hoped. After her visit, we followed the usual routine of working for the Diamond "K" cannery until returning to Puget Sound in early October. I quickly got a short job with the Libby Company, as an AB on the *Libby Main*. She was a small cargo ship, going north to bring back all the salmon canned for the company during the season. Following that three-week trip to Southeast Alaska, I went down to the union hall to take out a card for a job. This was in November of 1935, when shipping seemed to be picking up and jobs not so scarce as they had been.

Of the jobs available, the one that appealed most to me was on the *North Haven*. They were looking for six sailors for an interesting voyage, going across the Pacific to install bases for the Pan American clipper ships. I didn't like the fact that it would be a six-month trip, but it was a good job and would be a new experience. After talking it over with Gwen, I went back down to the union hall to throw my card in for that job.

Even though we were not due to sail until January, the crew was hired early to work days until sailing time, preparing the ship for the long voyage. Spending nights at home, getting a few dollars a day subsistence pay during that time, I enjoyed those several weeks before we departed.

This expedition of the *North Haven* was to be an historic voyage. As I had seen for myself, the Japanese were extending their military and commercial influence throughout the Pacific area, while the United States had limited its own defenses and influence in the area because of a policy of isolationism popular at the time. Concerned about this for years, the U.S. Navy finally saw a possibility of improving the situation through a commercial venture, riding the coattails of Pan American Airways as they established air service to the Far East with huge flying boats.

Juan Trippe, president of Pan American, already had a close relationship with the U.S. Government, winning support in establishing mail and passenger routes to South America. When he discovered U.S. Navy interest in his plans to fly the Pacific, he made contacts with the Navy that eventually reached into other parts of government all the way up to the White House. The government got quite involved with Trippe and his plans for sending clipper ships across the Pacific for many reasons, the most important probably being fortification of

An Historic Voyage

Pacific islands in preparation for a war with Japan.

The inaugural flight across the Pacific took place just about the time I went to work on the *North Haven*, on November 22, 1935. Preliminary work on the bases had been completed by that time, providing refueling and maintenance stations for the first flights carrying only mail. We would be completing the work and adding passenger facilities, getting ready for the first passenger flights beginning in October 1936.

The *North Haven* had an interesting background. She was built for the U.S. Shipping Board in Japan during World War I, when we were allies, named the *Eastern Gale*. After the war she was bought by the Booth Fisheries Company, renamed *Perry L. Smithers*, and used in their Alaska cannery operations for carrying crews north and bringing the salmon pack back to Seattle. Later sold to the Northland Transportation Company, she was again renamed, becoming the *North Haven*, and leased to a Swedish shipping firm which in turn chartered her to Pan American for this voyage. There was a lot of speculation about the roundabout way Pan American had chartered the *North Haven*, as U.S. Government interest in the airline's Pacific venture was an open secret.

With a big galley aft under the poop deck, and plenty of crew quarters, she was an ideal ship for this voyage across the Pacific with a construction crew. During the weeks before sailing we painted the ship, and cleaned out and painted the quarters. Lots of stuff in the galley and food locker was going to be thrown away. There was no objection when we asked if we could take some of it home, so I took a case of dried peaches up to the apartment.

"What can I do with that?" Gwen asked. "Why don't you make a peach pie?" I suggested. The pie she made was awful. She'd just made a crust and thrown the dried peaches in before baking it. I should have known. Homemakers weren't used to working with dried food the way seamen were. Dried food was used on the ships a lot then because refrigeration was limited.

The *North Haven* was in beautiful shape when we sailed on New Year's Day, 1936. Dave Girdwood was one of the dignitaries there to see us off. His Girdwood Shipping Company acted as agent for the Swedish firm chartering the ship to Pan American, further complicating the line of responsibility for the project. My shipmates were amazed when Dave broke loose from the group of bigshots to come over and shake my hand, wishing me luck. Dave was an old friend of mine from many years before, a very democratic fellow whose head wasn't turned by rising to own his own shipping agency.

Our first port of call was San Francisco, to load cargo for the trip. While laying there the firemen and oilers complained to the chief mate that they had no clothes lockers in their quarters. The mate came to me, asking if I was handy enough with tools to build a locker for them. I said I could do it if that was all right with the firemen and oilers, and with the ship's union delegate. I didn't want to get in trouble with the sailors' union.

They all agreed, so I started the job. I had all the framework installed and was making doors for the lockers when the carpenters in the construction crew learned what I was doing. They raised hell about it with the firemen and oilers, who agreed with them that as a sailor I shouldn't be doing that work. I then went to the mate. He just took me off the job and put me back on deck work. Those fellows did themselves in with that silly protest. The carpenters never got around to building the doors, so the lockers stood open for the entire voyage.

We took on quite a load there in San Francisco. A big barge and three launches were lashed down on the foredeck. The launches were twin-screw tugs about thirty feet long, nice little boats to be used for towing during construction at the islands. One was named the *Midway*, another the *Wake*, and the third the *Guam*. There was a lot of lumber loaded, three pre-fabricated hotels, much aviation gas carried in drums in the holds, and everything else imaginable that would be needed for the bases on the islands. The cargo even included large solar heating panels for the hotels, made up of copper coils with heavy black paper backings covered by glass.

The ship was heavily loaded when we finally pulled out of San Francisco Bay on a stormy January day, bound for Honolulu. The sixty-five men in the construction gang, who had their own quarters and used the big galley under the poop deck, suffered terribly from seasickness during the rough trip to the Hawaiian Islands. After taking on a few more items during our three or four days at Honolulu, we departed for Midway. There had been no shore leave given during our time in Honolulu. Those of us in the deck crew were to go three months without stepping ashore, from the time we left San Francisco to our arrival in Guam.

Arriving at Midway, we first put one of the small tugs in the water to be used for towing the barge in to shore. Discharging cargo was a slow process, taking several weeks altogether. Our deck crew would load the barge, which the tug would tow to a dock to be unloaded by the construction crew. Midway had been occupied for some time, from before the turn of the century, because there was a booster station there for the Transpacific Cable Company. The dock had been built for that facility, many tons of soil had been hauled in from the mainland, and pine trees were planted. The pine trees had grown to a good size, looking strange on one of the usually barren Pacific islands.

Another unusual sight at Midway was on Eastern Island, one of the three atolls making up the group. There were about a half-dozen burros living on that island and running wild. We were told they were left there after construction of the booster station for the cable company, nearly forty years earlier. Somehow they had survived and reproduced, though what they ate on that barren reef I couldn't imagine.

Offloading was more of a challenge than might be imagined. The three coral atolls making up Midway surround a lagoon. I believe the lagoon had been dredged since we were there in 1936, but at that time it was impossible to take a ship the size of the *North Haven* into the lagoon. Because of that we anchored in the lee of the atolls when weather permitted, and when the swells were too great to anchor we'd drift in the lee or run slowly to stay in a comfortable position. Midway was the first stop for the Pan American clipper ships after leaving Honolulu. When work was completed there we sailed on to Wake Island, the next stop for the big flying boats as they hopped across the Pacific. The little tug *Midway* was left at the island for which it had been named, along with the barge carried from San Francisco and used for offloading there. With more room on deck now, the carpenters set to work during the several days passage to Wake Island, building a barge to replace the one left at Midway. The second small tug, *Wake*, would be left at that island along with the barge built during this trip, and the third tug left at the island for which it was named, Guam. It was a well-planned operation.

Unlike Midway, Wake Island had been entirely unoccupied when the first Pan American expedition arrived there. They found the island infested with thousands of rats, apparently the only survivors of a shipwreck many years earlier. By the time we arrived the workers had baited oil drums with the tops removed, putting ramps up to the drums in such a way that the rats would fall in when they went up the ramps. They collected hundreds of them at a time in the barrels, then poured gas into the barrels and burned the rats. It was the only way they could get rid of all those rats to make the place habitable.

At Wake Island we were unable to anchor, having to lie about a half-mile offshore in deep water during all the weeks of offloading. This meant keeping up steam all that time, drifting and sometimes under way to shift position, but we had no problems with the ship itself.

The barge used to haul supplies to the island was another matter. A kedge anchor was used to snub the barge through the surf to the beach, but the kedge didn't want to hold in the coral. There was a native Hawaiian in the construction crew who was a powder monkey. He came up with a bright idea to hold that kedge in the coral: to tie several sticks of dynamite to the anchor before lowering it to the bottom. His theory was that setting off the dynamite would blast a hole into which the kedge would drop. His idea caused him a lot of teasing for the rest of the voyage, as the blast when he tried out his theory simply blew both flukes off the anchor.

It was a pleasant and profitable time for our crew at Wake. The weather was warm and comfortable, while the work was both interesting and rewarding. The deck crew handled offloading of the ship, for which we got extra pay. That included the turnaround time for the barge at the island, amounting to about an hour and a half while we waited for the tug to pull it back for another load. In addition to that work we stood regular watches at night, since we couldn't anchor and had to stand off the island either steaming slowly or drifting.

The two-man crew of the tug Wake slept aboard at night while we were offshore on the *North Haven*. One morning when we moved in close to the island about dawn, the tug blew up. The twin engines were run by gas, and fumes had accumulated when they went to start the engines. The man close to the engines was blown into the flames, burning him terribly, while the second man was not hurt.

Seeing the explosion, we put our lifeboats over as quickly as we could and rushed to give them assistance. When a hundred feet from the boat we could see they were in a terrible predicament. The heat was so intense that we could feel it from that distance. The two men, who were on the bow of the tug, wanted to jump into the water, but didn't dare because the flames had attracted a lot of sharks around the boat. We managed to get right up to the bow and take them off.

Fortunately, the *North Haven* had a doctor and a ship's hospital. Dr. Bill Davidson was a very nice fellow who had been with us all the way from Seattle. He took care of the badly burned man as best he could. The poor man was in so much pain and suffering from his burns that he went out of his head. I found him wandering around deck one day, on feet with flesh almost all burned off. We got him back to the hospital room, and the stewards department people kept an eye on him from then on. Under the circumstances there wasn't anything we could do with him until we got to Guam, where he was placed in a hospital ashore.

The tug *Wake* was a total loss. To replace her, the *Guam*, destined for the island of that name, was put overboard and placed in service. Being short the tug operator who was burned, and knowing that I had a lot of towboat experience, they offered me the job. I said I'd take it if they'd pay me right. I was making $60 a month as a seaman, plus good pay for the cargo work. My offer to run the tug was $10 a day plus the same cargo pay as the other seamen would be receiving.

They thought that was too much, choosing another fellow with little experience. He ran the *Guam* on the beach during his first trip, holing and sinking her. When they got her raised, patched and put back into service, I couldn't resist rubbing it in by telling the supervisors that that wouldn't have happened if they'd hired me.

Despite all the troubles we managed to complete offloading at Wake after another several weeks, and sailed on to Guam. That island was considerably more developed than the others, with a good harbor, a naval and air base at Piti, and a town about six miles away named Agana. With the ship anchored in the harbor near the naval base, discharging the third hotel and other cargo went much faster than at the other islands.

Japanese interest in our operation and other U.S. activities in the area was made obvious by their scouting planes flying over us nightly. Guam lay in the middle of the islands of Micronesia, which had been placed under Japan's administrative jurisdiction by the League of Nations as the Japanese Mandate. Ignoring terms of the mandate, Japan had by this time established air bases and fortifications on several of the islands. With Saipan only 115 miles from Guam, the Japanese had no problem in surveying activities there both by air and sea.

We had seen Japanese interest in our activities once before, far removed from the Japanese bases. That was at Wake Island. In the dark close to dawn one morning there, while I was standing watch with the second mate on the bridge, I saw a big hulk close to us with no lights. The mate turned our searchlight on and there we saw a big Japanese cruiser skulking by us with no running lights, obviously spying on our operation. It was another reminder that the situation between the U.S. and Japan in the Pacific was getting tense.

We lay at Guam about six days to complete the work, and the crew finally got a chance to go ashore. I remember going through a beautiful waterway up to that lovely town of Agana, where we got a few "refreshments." We didn't get wild, though, most of us returning to the ship in an orderly fashion. But when it came to sailing time the bosun and one other man turned up missing. A search was started for them ashore, using a detachment of marines. After about six hours they were found hiding in a sugar field. It turned out that the bosun had gotten himself drunk, decided that Guam was a fine place to stay, and had talked the other man into jumping ship along with him. When the marines returned the two men to the ship, we sailed for the Philippines with a full crew.

It was a beautiful trip to Manila, running at about ten degrees latitude north in the tropical sun. Everybody had been living in nothing but shorts during most of the time since leaving Honolulu, so by now we were all deeply tanned. Going through Sebastian Straits, we followed a passage through the inland sea dotted with islands covered in tropical foliage. The *North Haven* anchored in the harbor at Manila, laying there several days to discharge the remaining cargo for the Pan American Station. Manila was the next-to-last stop on the clipper ship route. Juan Trippe had tried to negotiate landing rights in China itself, or Hong Kong, without success. He finally settled for Macao, a Portuguese possession not far from Hong Kong and Canton, which would be the final stop and turn-around point for the clipper flights.

The crew was able to go ashore at Manila. Taking a boat in to Legaspi Landing, we roamed around the city and had a few highballs, as sailors like to do. Late one night we wandered into Manila's walled city, taking a cab through there to see the sights. We were told later that it was only luck that kept us from getting into trouble, because that was a tough area of the city. One crew member, a Norwegian named Ole, didn't have such luck. He got drunk, was given some knockout drops, and woke up stark naked in a ditch. The police put a coat on Ole and brought him back to the ship, a lot poorer but not too much worse for the wear.

While laying in Manila a wire was received, informing us that the dock at Midway had been washed out by a big storm. We loaded planks and timbers to repair the dock on the return trip, lumber that was all teak wood. I was amazed to see such valuable wood used this way. When I mentioned that to Dan Vusovich, he told me that it cost no more than fir planking would cost in the Pacific Northwest. Dan was supercargo on the *North Haven*, the officer in charge of cargo and operations. A native Slovenian who spoke what we called "good broken English," Dan had one time been a banker in Fairbanks. He was a wonderful fellow with a personality that helped make the voyage pleasant for the crew.

Leaving Manila we steamed directly to Wake to put a few things ashore there, then on to Midway to repair the dock. While there, using the same procedure as before with a tug and barge taking the lumber ashore, we did some anchoring. The weather had been good when we anchored, but a small gale came along causing the ship to surge up and down in the swells. The captain's worries about those swells during our stop at Midway on the way out proved to be well-founded. The chain parted and we lost the anchor. With the gale still blowing we then moved closer into the lee of the island to anchor again in 17 or 18 fathoms of water, only to have the chain break again with the loss of our second anchor. The water was so clear that we could see the anchor and chain down there, laying across the coral, but there was no way we could retrieve it.

With two spare anchors and some chain left, we tried again with the same luck. In the next days the *North Haven* lost both those spare anchors. It was lucky that Pan American had an old-fashioned stock anchor ashore, a small one that weighed only about 1500 pounds compared to the 4500-pound anchors we'd lost. We brought that out from shore, and with the very little chain still left were able to use that for the rest of our stay at Midway, somehow managing not to lose it.

One of the sailors had picked up an unusual venereal disease in Manila that had him in terrible shape. Dr. Davidson didn't have the medicine and equipment to treat the poor man, who was in agony, so it was decided to evacuate him on a clipper ship that was due to arrive, bound westward for Honolulu. A message was sent by radio from the ship to the Pan American station, asking them to send a launch out to the *North Haven* to pick up the sick man. The seas were breaking so hard on the reef at the time that the station said they couldn't get a launch out to us.

The man's situation was so urgent that a call was put out for volunteers to row him to shore. I volunteered to go along with seven other oarsmen, the second mate steering and in charge, the patient and the doctor. After about an hour of rowing we reached the mouth of the lagoon, where the breakers looked very dangerous. The second mate was an old Finn I'd shipped with before on the *Oduna*, Mr. Kilander. "Oh, we'll make it.", he said. I didn't think we would make it once we entered those breakers. Watching the mate haul on that steering oar, which bent so much at times I was sure it would break, I knew we would swamp and all be lost. But somehow we did make it and got the patient to the Pan American station to wait for the plane the next morning.

"Well," said Mr. Kilander, "now we go back."

"The hell we will!" we all said at once. There was no way we were going to pull back through that surf on the way out, an even more dangerous trip than it had been coming in. "Yeah, you will," Mr. Kilander said, "otherwise it's mutiny." "Okay," I told him, "it's mutiny."

At that point Mr. Kilander went up to the Pan American station and wired out to the *North Haven* that the crew refused to return to the ship. The captain radioed back: "Don't you dare try to come back through that surf. Wait there until morning." We did, spending the night rolled up in tarpaulins in the warehouse. Although it was warm during the days on the island it was cold at night, so our one overnight visit to the island wasn't spent very comfortably. The weather had moderated so much in the morning that the Pan American launch was able to tow us back to the ship, saving us a lot of rowing.

One bit of luck turned up on the next-to-last day before leaving Midway. Every time we used that anchor from the Pan American station after losing our own we anchored over the site of the first anchor we'd lost, in hope that the anchor would catch on the lost anchor chain. The luck on that day was that we actually did hook the lost chain, raising it to the surface with the borrowed anchor. Mr. Kilander took me and some other sailors with him in the work boat up to the bow of the *North Haven*, to attempt placing a shackle through one of the lost chain's links so we could retrieve it. It wasn't an easy job with the ship surging a lot in the swells, but we somehow managed. We then ran the chain up through the empty hawser and pulled it in with the winch, another tough job because the shackles were running backwards. We got it done, though, and the captain was happy that he could return to home port with at least one anchor of his own.

After a couple of weeks at Midway we departed for home, calling at Honolulu before reaching the States. With little for the crew to do while in port there, we were all ready for a celebration of the work we'd done. I decided to party with one of my friends in the crew, Louis Lindstad, a husky AB who was a football player at the University of Washington. The first thing we did was to go for a haircut, getting ourselves spiffed up before looking for a bar.

We were standing on a street corner waiting for a cab when a Kanaka drove up in one, got out, and accused us of not paying the fare. We'd walked that far from the ship and certainly not used his cab, but he started arguing with us in a loud voice. Louis got so mad that he hit the Kanaka, knocking him out into the street. The police showed up, arrested us, and took us to jail. It was obviously a racket, but there wasn't anything we could do about it. We were charged and thrown into a cell for many hours. Somehow we finally got word to the ship just about the time they were getting ready to sail. Thank heavens for Dan Vusovich! He refused to let them sail, came up to the jail, and saw us. "It'll cost you five dollars each to get out of here," he told us. "Pay it!", Louis and I told him. "You're also going to have to pay the cab fare," Dan said. That was harder to swallow, but Louis and I decided to pay it rather than fight the charges.

If the visit to Honolulu was a terrible disappointment to Louis and me, our arrest was a good thing for the ship. When we finally got back to the *North Haven* we found a drunken and disabled crew. The captain was drunk. The chief mate was drunk. The winch driver was drunk. All the sailors were drunk except Louis and me. The only crewmen still functioning normally besides the two of us were the second and third mates.

All these drunks were getting ready for departure of the ship. One of the last operations of a departure is to bring the gangway in. A sling is placed under the gangway to lift it aboard with a winch, the sailors untie it where it is made fast to the ship's rail, and the chief mate is last to come aboard from the dock, where he has been supervising the operation. In this case the sailors forgot to untie the gangway from the rail, and the winch driver put the winch into gear just as the mate stepped on the end of the gangway. With the gangway still fast to the rail, the dock end flew up and catapulted the mate into the air. The mate, Mr. Peterson, did a full backwards somersault, landing on his feet with his pipe still clenched in his teeth, standing there as if nothing had happened.

The second mate sent Louis to take over as winch driver, I saw to it that the gangway was unfastened from the rail, we pulled Mr. Peterson aboard and got away from the dock. The next problem was standing watches. With the second and third mates the only officers on their feet, and Louis and I the only sailors, I took first watch with Mr. Kilander and Louis took the second watch with the third mate. It was a strange situation, with no lookout or anybody on deck. After several hours of steaming that way, Mr. Kilander relieved me at the wheel and I went below to see if I could find anybody to give us a little help at the wheel. I finally found a refrigerator mechanic who was sober, a nice fellow who liked to steer, and he came up to the bridge to give us a hand. The rest of the crew was beginning to come around by the end of the second watch, but they were a sad bunch for the next couple of days.

Lightly loaded as the ship was, we rolled all the way across the Pacific to San Francisco, the last stop before returning to Puget Sound. We were there just for a couple of days, unloading a few remaining items at that port. While lying in port the bosun quit. Joe King, the winch driver, was given the bosun's job. A couple of sailors also left the ship, so we took on another two sailors new to the crew. One of them was a young fellow from the University of California at Berkeley. As the crew was having coffee in the crew's mess one morning, the college student was giving a big talk about all the blessings of socialism, and the faults of our system. Joe called the sailors out on deck, but the new sailor kept talking while the others listened. "All right," Joe yelled at them in his Irish brogue when he came in the galley again, "if you're going to shairr the wealth you're going to shairr the work! Get the hell out there on deck." Knowing Joe, and the way of bosuns, I'm sure that kid got the dirtiest jobs on deck for the rest of the time he was aboard.

It was July when we got back to Seattle, the end of the trip and a happy time for me. The crew was paid off. Having been on the *North Haven* from November 1935 to July 1936, receiving extra pay for cargo work, it was the largest check I had yet seen. And it was good to be back with Gwen. Our love and trust for each other had grown so much through the years that we took that check and some of her money, and opened a joint banking account. It began to look like we would be married after all.

CHAPTER TWELVE

Endings and Beginnings

THE NEXT TWO YEARS of my life were to see endings and beginnings: the ending of my bachelorhood and beginning of a marriage, the ending of working with Dad on the *Hannah C.* and the beginning of a full-time deep sea career, the ending of working on deck and the beginning of serving on the bridge as an officer.

My arrival back home in July 1936 came just a few days before Dad was leaving for the usual late summer work in Alaska. Paul and I went with him once again, returning early this year around September 10th.

I hurried up to Ben Tipp Jewelers to see Gwen. The first thing she said to me was "We're going to get married!" I had written to let her know the date of our getting back from Alaska, and she already had our friends Johnny and Nelda Walker all lined up to go with us down to the courthouse to get a marriage license.

"How come we're getting married now?" I asked her.

"You've always wanted to get married. Come on, let's go."

She whisked me out of the store, we met Johnny and Nelda, and before I knew it were at the courthouse. Coming up directly from the boat to see Gwen, not planning anything like this, I was embarrassed to find that I didn't have the two dollars and fifty cents for a license. That tickled Johnny Walker. "Well, if Holger and Gwen are finally going to get married, I'll sure as heck dig up the money for the license." He did, and I took a lot of ribbing for that even though I paid him back as soon as I drew my summer's salary from Dad.

Gwen really did have everything arranged. The wedding date was set for just a couple of days after I got back from Alaska, September 12th. Besides our families and some other friends, the guest list included the Walkers, Dave Girdwood, the captain and an agent for a Danish ship in port, all of them stopping by on the way to a party.

We were married by a minister at the home of Gwen's sister and her husband, Dorothy and Art Sherman. The minister went home right after the ceremony, and my parents left not long after that. My mother being an abstainer, we held off opening the keg until they were gone, then had ourselves quite a celebration.

Our honeymoon was spent in California. After returning in late September, moving into Gwen's place at the Lexington Apartments, I went down to the union hall to take out a card. But things weren't looking too good on the waterfront. It seemed as if the shipping companies were testing union power following the strike of 1934. The tensions brought another strike on October 29, 1936. This one lasted until February 5, 1937.

Unlike the 1934 strike, which took place in summer when I could work with Dad, this one came during a season when Dad wasn't so busy. I was lucky to be in fair shape to weather that time; having come off the *North Haven* trip and the late summer on the *Hannah C.*, with Gwen working and our used car business on the side, we weren't hurting for money. Most of the sailors were, though, and still without strike pay or other unemployment benefits. The soup kitchens went back up, with long lines of sailors forming to get some food, and picket lines were organized. I joined the picket line two or three times a week, donated what I could to the cause, and once in awhile was able to go with Dad on a powder run for Du Pont to earn a few dollars. The only shipping moving at all was an occasional emergency run allowed to take supplies to communities in Alaska running short on food and other necessary items.

The union hall was jammed when I went down there to take out my card at the end of the strike. There was a lot of excitement and activity as the ships tied up for so long were getting ready to go back to sea. The job that interested me was on the *Silverado*, a coastal ship under the Moore McCormack flag. She was a small flush-deck ship with three hatches, all forward of the bridge, carrying lumber and other goods between Puget Sound and California ports. A schedule like that would give me more time at home.

Throwing in my card for an AB job on the *Silverado*, I joined her at the Pier D Moore McCormack dock while the cargo was being worked. Under terms of the strike settlement the deck crew on the ship worked cargo on one hatch, while the other two were worked by longshoremen. Not only did we get extra pay for the cargo work, but we got overtime as well. And the food was good!

Endings and Beginnings

After taking on general cargo and a full 17-foot deck load of creosoted piling we sailed to Monterey to discharge, went to San Francisco to take on cargo for the north, and headed back to Puget Sound. On the Sound we loaded at a number of ports, some of which are no longer active and hardly recognizable as sawmill towns. Besides general cargo we would load lumber in Tacoma, and almost every trip called for lumber at Port Ludlow and Port Gamble, both more like resort towns nowadays than like the old mill towns they were.

With a good deck crew, good officers, and a fine captain, it was an excellent job. The captain was a Finnish fellow named Seilo, who had pilot papers for all the ports we called at, a quiet man and good skipper. About the only discomfort was the living quarters, a focsle with three-tiered berths. As the newest man aboard I got the topmost berth. That wasn't as much of a problem, though, as getting back and forth to the focsle with that 17-foot high load of pilings on deck. In heavy seas it took some doing to climb up over those black pilings in the darkness at night, slippery from spray, to travel all the way aft to the bridge. There was a lifeline to follow, but it was still a trick to negotiate that trip at every change of watch.

My partner on the *Silverado* was a nice chap named Al Hammond, who was born in Norway and had come with his folks to Seattle while still quite young. He and I got along so well that when I told him about my summers with Dad on the berry run and in Alaska he wondered if he might get a job with Dad. We signed off the *Silverado* together in late May, taking a few days off before going back to work. Dad liked Al, too, so he hired us both.

The summer of 1937 was an especially memorable one for me. Al stayed with us all that season through the berry hauling and including the trip to Alaska. He was a good addition to the crew, helping to make it an especially good summer. There was always some fun besides the hard work. One of the things we used to do up north when we were laid up on the outside of the islands, waiting for fish to fill a trap, was to go hunting. I remember that one of our hunting trips that summer was on Kuiu Island, where Al and I climbed a mountain about 1500 feet high. It was a tough climb, worth it because at the top we found a herd of deer. Using a .30-.30 rifle always carried on the boat, I picked off a beautiful buck. It was hard getting it down to the beach. We rolled it down the steep hillside, floated it down a small river, then cleaned it out on the beach while Paul came in from the *Hannah C.* by skiff to pick us up with a new supply of meat for the crew. I guess hunting that way wasn't quite legal, but we'd do it every once in awhile for the meat it provided.

Al was a good accordion player, carrying his squeeze box with him wherever he went. Another memory of that summer is of sitting around on the boat with Al playing while the rest of us tried to sing along with him. Except for Paul none of us had a good voice, but we all enjoyed those songfests.

One of my favorite runs with the *Hannah C.* in Alaska was up near Rocky Pass, above Straits Island and Sumner Strait. We anchored off Montecarlo Island, meeting the gillnetters in the area to load their fish for the cannery. Loading two boats at a time, one on either side of the *Hannah C.*, the crew took turns acting as tallymen to count and record the fish from each boat. At that time of year they were mostly pink salmon, or "humpies" as they're called there, with fewer silvers and an occasional big king. While at Montecarlo Island we often went ashore to visit a fox rancher and his family, and wander around watching the foxes. The only way to get ashore was by skiff, as there was no dock there. We always had to be careful not to leave anything in the boat, and to turn it upside down on the beach, as the foxes otherwise would steal loose items and leave their droppings on the seats.

Once in awhile we carried passengers along with us. Two of them during that summer of 1937 were a Mr. Cunningham and a Mr. Dahl, the first the owner of the Wrangell liquor store and the last the owner of the telephone company there. Both good friends of Carl Thiel, owner of the cannery, they arranged through him to get a return ride with us from Warm Springs Bay, where they had been spending a couple of weeks on vacation. Warm Springs Bay was a pleasant spot for that, with a hotel and inn located on a pretty little harbor, a place where many of the fishermen went to take a rest.

After delivering supplies to the watchman at the Kings Mill trap we ran to Warm Springs to pick up the two men, coming in about 4:30 AM. Wandering through the village we tried to find where our passengers were staying, with no luck. Paul, Al and I then walked up to a dam holding back water for the drinking supply of the town, a beautiful spot in the early morning light. On our way back through town, Paul looked up at a sign on a store: "Liquors and Tavern." "Maybe they've got rooms here," he said. He was right. We knocked at the door. After a time a lady came to ask what we wanted so early in the morning. Yes, Mr. Cunningham and Mr. Dahl were staying there. She woke them up and we got them out to the boat, taking them back with us to Kings Mill to brail the trap before returning to Wrangell.

Little incidents like that helped round out another good summer in Alaska. Al left us at the end of the season. Like me, he went on to become a deck officer and eventually became master of one of the American Mail Line ships. In later years he worked as a Columbia River bar pilot and by now, I suppose, is retired.

At the end of that fine season with Dad, I shipped on the *Margaret Schafer*, belonging to the Schafer Brothers Shipping Company of Aberdeen. When I joined her in early October 1937, she was loading at the Nettleton Mill in Seattle and needed a couple of sailors. We sailed from Seattle to San Pedro with a load of lumber, not the normal route for this ship as she usually carried lumber from Aberdeen and Willipa Harbor to California ports. Those late fall trips were pretty rough from the standpoint of weather. Crossing the Grays Harbor bar was quite dangerous at times, and running down the coast in fall storms gave us a beating. We almost lost our forward deckload of lumber in one of those storms, coming out of it with the lumber looking more like a haystack than a proper load.

But I liked that job. When we came into port at Aberdeen, Gwen often met me and if on a weekend we'd drive to Seattle to spend the time together at home there. The master, Captain Thor Berg, was a very nice fellow.

One of the winch drivers on the ship was a man named Rosen, who lived in Aberdeen with his wife. The other winch driver was called "Teddy," a Dane from my mother's home island of Bornholm. Teddy was quite a character. He'd swipe anything loose on the ship and sell it ashore. I'm sure he'd have sold the anchor if he could have carried it.

While we were loading and preparing to leave Aberdeen one winter day Rosen invited me, Teddy, and another friend on the crew for lunch at his home. We took off at noon and drove up to his place, where Mrs. Rosen had fixed a wonderful duck dinner for us. It was a great meal. Afterwards we returned to the ship, finished the preparations, and headed out over the bar for San Pedro.

Captain Berg called for the steward as we neared San Pedro, asking him to go down to the freezer and get out the ducks he had put there. The captain explained that he'd shot the ducks on a hunting trip while home, and was taking them to a friend in San Pedro. The steward left to fetch the ducks, then came back to report that there were no ducks to be found there. Captain Berg went down to the freezer to look for himself, with the same results. It turned out that the duck dinner we'd enjoyed at the Rosen's was courtesy of Teddy, who had stolen the captain's ducks from the freezer and given them to Mrs. Rosen to prepare. I'm sure the captain knew who had swiped his ducks, but he was a quiet man who accepted the loss without making a big fuss over it.

On some runs the *Margaret Schafer* called at San Francisco to deliver lumber. During one of those calls in December, 1937, I was helping secure the booms to get ready for returning to sea. The ship had a slight list as a couple of other crew members and I pushed against the long, heavy boom, trying to get it in place for lowering into the cradle. It was so heavy that I put one foot on the cargo winch to give myself a purchase. As I did that, another fellow jumped up to give us a hand. In jumping his foot accidentally hit the throttle of the winch, pushing it to full speed. I fell down into the middle of the speeding winch, which banged me around something awful. It was lucky that the winch was unwinding, or I certainly would have been killed.

Somebody stopped the winch as quickly as possible and got me out of there, carrying me back to my bunk, barely conscious. The ship was already under way for San Pedro when this happened, a trip of about a day and a half. I was terribly sore all over, staying in my bunk until we reached port and then going up into the city to see a doctor. I was black and blue from my neck to my heels, with the worst pain in my tail bone. X-rays showed nothing broken except for the tail bone, which was banged up. The doctor offered to put me in the hospital there in San Pedro, but I preferred to return to Seattle. He gave me some painkillers, which helped me get through the return trip of the *Margaret Schafer* to Grays Harbor, laying in my bunk all the while, and on the car trip with Gwen back to Seattle.

The company gave me a slip to go to Dr. Buckner, a bone specialist in Seattle. After x-raying me again, he prescribed a treatment which was a series of shots given once a week, a "bone-building" injection of something that I could taste in my teeth every time I got it. Being laid up because of the accident and having to go through the series of shots, I got $15 a week in compensation. Still able to move around, having nothing else to do, I went down and enrolled in Captain Kildahl's Navigation School with the idea of getting a mate's license. It was early March, 1938 when I sat for a third mate's license, passed the examinations, and received a ticket as a third mate. It could be said that I received that ticket "by accident," as it was due to falling into the winch on the *Margaret Schafer* that I ended up going to navigation school and sitting for the license.

In the meantime, I had been dealing with an attorney in the firm of Bogle, Bogle, and Gates to settle with the shipping company on the accident. Their first offer was a settlement of $200, which I refused. After some negotiations I finally received $900, just about the time I received my third mate license. By then I was feeling quite well, had come out all right financially, and had my new ticket. I went right down and joined the Master Mates and Pilots Union, looking for a job on the bridge.

After waiting some weeks for that kind of position I went back to the Sailors Union, where I had kept my membership. There I learned that they were hauling the *Baranof* out of the boneyard and needed three quartermasters. A friend named Joe Bington, another fellow, and I all threw our cards in at the union and got those positions.

A quartermaster is a petty officer who handles the steering on a ship, wearing a dark suit with a special cap rather than the dungarees worn by a seaman. On the *Baranof* we stood four-hour watches, with half an hour off in the middle of each watch to get a cup of coffee. My watch was with a pilot named Hansen. "Small-Eighth" Hansen, he was called, because of his steering orders. In those days we steered by compass points rather than degrees. This pilot was so exact that he would break the points even smaller than those usually given. He'd call out something like "Northwest one-eighth north — and a small eighth," as if we were following a railway track and could actually steer that close. I'd just repeat the order and steer as close as I could to the course I thought we should be on.

That turned out to be a good job. We went way out to the westward parts of Alaska, calling at a lot of small ports and working cargo so much that we got lots of overtime. The quartermasters also worked cargo, and so I joined with the rest of the crew when we stopped at Ketchikan on the return trip, filling our number three hold with frozen halibut. We got under way again as soon as the load had been taken on, the crew moving the halibut into the cold storage lockers by hand. It was a long, hard job, and I found myself taking full four-hour steering watches in between periods of carrying those heavy fish from the hold to the cold storage.

By the time we got to Seattle I was hurting so bad in my ankles that I could hardly walk. I went to a doctor who ordered me to stay off my feet for awhile to let them heal. That job on the *Baranof* was so good that I hated to leave it, but my feet were in such bad shape that I had no choice. Once again I was in an idle period forced by physical problems brought on by my job.

While I was resting up, Gwen and I decided to look for a house. We had been at the Lexington Apartments for many years and were beginning to get tired of that kind of living. Gwen finally found one she liked out in the Laurelhurst area, a nice six-room place that suited us just fine, a Tudor-style brick home on a corner lot. We paid $5,250 dollars for the place, moved in, and found that we were happier there than we had ever been.

By now it was May, 1938. I had been putting in applications for a third mate's job with no luck. So I went back with Dad again on the *Hannah C.*, working the strawberry season and going north with him on the contract with the Diamond "K" Company in Wrangell. The season was no different than before, but it turned out to be the last I'd ever spend with Dad. Considering that I had started out with him as a boy, and spent most of my summers working on *La Blanca* and the *Hannah C.*, I'd been with Dad for about a quarter of a century. It was quite a turning point in my career, the end of an era.

We came down from Alaska early again this year, in September. I no sooner got home than a call came from the Griffiths Company, asking me to come to the office to see Mr. Wuthenow. They were putting *El Cedro* back in service after being laid up for some time. The third mate's position was open. I was happy to take the job, and was asked to report to the captain the next morning. When I told Gwen about it she was both happy and sad; happy for me that I would finally be sailing as an officer, and sad that I would be gone again. We were both happy that I would now be getting an officer's pay: $125 a month base, plus overtime for supervising cargo handling.

El Cedro had been laid up in Winslow. When I took the ferry over there the next morning I found the other two mates already aboard, Chief Mate Perry Cedar and Second Mate Ernest Langstrom, both of them men with master's papers. It was common in those days to have captains sailing as mates, as there weren't enough skipper's jobs for all those who held master's licenses. The captain of *El Cedro* was in Seattle on ship's business that day, so we got steam up and Perry Cedar took the ship over to Seattle's East Waterway to pick up some gear. While we were crossing the sound I got out my brand new license and hung it up with the others in the pilot house. All of the three other officers had old, well-used master's licenses with several issues and many piloting endorsements.

"Now you're going to see a real license," I said as I put my new one alongside theirs.

Ernie Langstrom grinned. "That's all right," he said, "everybody has to start that way." He was a fine old gentleman.

We picked up the skipper, Einar Ericksen, and went down to Tacoma to take on lumber at the St. Paul mill there, a full load destined for San Pedro. It was there that my duties as third mate began. The chief mate was in charge of all loading of cargo, with the second supervising loading of the after deck and the third mate responsible for the forward deck. We saw to it that the lumber was stowed right, marking the various lots with different color paint stripes when the supercargo ashore told us the lots were changing.

My bridge watch came as we were sailing up Puget Sound. Although I didn't have pilotage papers at that time, the captain knew that I was very well acquainted with the Sound and he left me alone on the bridge, in charge of the ship. I had spent so much time in wheel houses by then that I felt like an old-timer. Leaving the Strait of Juan de Fuca, we set a course towards San Pedro that would take us several miles off the long reef that runs out from Cape Blanco.

I got used to the duties of a third mate very quickly. While docking or anchoring my position was at the engine room telegraph, sending the orders for power changes such as "half astern," "full ahead," or "stop engines," keeping a log book to record all those changes. My bridge watch was from 8:00 PM to midnight, and from 8:00 AM to noon, a watch I liked very much because I could get a decent night's sleep from midnight until the morning watch. I would get up in time to have breakfast before going up to the bridge to take my watch. My first duty there was to wind both the chronometer and the ship's clock, then I'd check over the log to see what had happened during the night.

In San Pedro we went to several docks to discharge the different lots of lumber, then to the Union Oil dock to take on oil. *El Cedro's* number two hold was really a deep tank, and she also had wing tanks in addition to the double-bottom tanks used for her own fuel supply, so we were able to take on about 15,000 barrels of oil as cargo for Puget Sound on the return trip. That gave us a good ballast for the trip up the coast.

We were on a regular lumber run from Puget Sound to San Pedro. *El Cedro* took about two million board feet of lumber, stowed both in the holds and on deck. With holds full, and a well-trimmed deck load of lumber, she rode fine in the sea on the trip down to San Pedro, rolling easy with nothing quick about the roll.

I liked that job on *El Cedro*. With my own quarters, eating in the dining saloon, and fine officers, it was a good beginning of my time on the bridge. In Puget Sound we sometimes loaded at Tacoma, sometimes at Everett and Bellingham, but most often at Olympia. We would take about a week to load, as no overtime was allowed due to lumber not being a very lucrative cargo. Because of that I was able to get home most nights to see Gwen, driving my car and taking along the captain and chief mate who both lived in Seattle. It worked out well for all of us.

This was during the late years of the Depression, when jobs were still pretty scarce. The deck officers and engineers stayed in their jobs with the Griffiths Company, since it would be hard to get back with the company if they quit. The company didn't mind our taking off for a run or two, though, if it was for the purpose of advancement. Towards the end of 1939 I did that in order to study and sit for my second mate's license, having served enough time then to qualify. When I returned to the ship I still was third mate, with Ernie Langstrom remaining in the second mate's position.

It was just about that time that things began to change in the shipping business. Lots of government contracts were being granted for shipping as World War II was approaching. We had stayed in the coastal trade with *El Cedro* through 1938 and 1939, but in early 1940 we started inter-coastal runs. The first of these was a load of lumber taken on in Olympia, destined for Camden, New Jersey.

Our first stop was at San Pedro, to take on fuel for the long trip. In our coastwise trade I hadn't had much chance to practice the celestial navigation I had studied so much for my licenses, since we navigated up and down the coast mostly by taking bearings off the points. Now, on the two-week trip from San Pedro to the Panama Canal, and again up the East Coast, I finally had a chance to put all that knowledge to work.

After the long trip to Camden we were kept on the East Coast for awhile. From Camden we went to Richmond, Virginia, to take on a full load of pilings for Cocosola, on the east side of the Panama Canal. Then it was up to Cuba to load sugar for New Orleans. Several months were spent like this, shuttling cargo around the East Coast, before we returned to Puget Sound.

Ernie Langstrom had left the ship before we started this trip to the East Coast, taking a chief mate's job on another Griffiths ship in the West Coast trade. The mate who replaced him was Harry Christensen, known as "Two-Step Harry." Legend had it that he got his nickname during the years he was working in the Alaska cod fishery. Coming back to Seattle with a big paycheck at the end of each season, he'd go up to the jitney dances popular then, where a man would buy a ticket for ten cents to dance with one of the girls there. Harry loved to do that. His favorite dance was the two-step. He'd dance the two-step all night long, winning a nickname that stuck with him all his life.

Old Two-Step was an odd character, a short, stocky fellow who was gruff and cranky at times but with a sense of humor and a good heart. An incident typical of his personality took place while we were lying in port one night, Two-Step tried to sleep in his quarters while the chief engineer was doing carpentry in his own quarters next door, banging away on some shelving he was installing. Harry didn't say anything about not being able to sleep for all the racket, but next day he put up a big sign on the chief's door: "Wood butchers wanted in Bremerton. Anything goes."

While we were loading oyster shells in Gulfport, Mississipi, for the return trip to the West Coast, Two-Step and I went up into town with some others of the crew to see the sights and hoist a few. We took a cab back to the ship, a trip of two miles or so. "How much do I owe you?" Harry asked when we pulled up to the ship. "Twenty-five cents," the cab driver said. Harry stared at him, saying "You bunch of guys are highway robbers down here." The poor cab driver started pleading with him. "Mister, I just can't do it for less than that. Times are still tough down here in the South." Harry grinned then, and gave him 50 cents.

Part of the load being taken back to Puget Sound was new Dodge cars, which filled the two after holds of *El Cedro*. Two-Step was on watch one night after we had transited the Panama Canal and were running north off Central America, when the horn on one of those cars shorted out and started blowing intermittently. The sound of the horn coming up through the ventilator shaft made Harry think it was a boat whistle. He went wild because he couldn't see any boats nearby, running out on the bridge to check with the lookout. "Where the hell are the lights? Do you see any lights?" The horn stopped blowing then, Harry went back to his station on the bridge, and then a few minutes later the horn started up again. By now Harry was so excited that he was doing a two-step up there on the bridge, running back and forth to try to spot the boat blowing its whistle, checking with the lookout, thinking he was in the middle of a tuna fishing fleet not showing their lights.

The lookout and the sailor on the wheel quickly figured out what was going on, but wouldn't tell Two-Step. When the chief mate relieved him, Harry told him about all the excitement during his watch. The chief mate realized at once what had happened. He had a sailor come roust me out of my bunk to go down to the hold and fix that horn, figuring I knew all about such things because of being in the used car business. I had a heck of a time getting under the hood of the car because these new models had locks on the hoods, but I finally managed to get in there and yank out the wires to the horn.

We got back to Puget Sound safely despite Two-Step's phantom tuna fleet, and then took two loads of lumber down to San Pedro. When that second load was discharged in mid-September of 1940 we took on a full load of cement for delivery to Balboa and Cristobal. Since we had to transit the Panama Canal to discharge the balance of the load at Cristobal, we then went back to shuttling cargoes around the East Coast on runs much like those before. Our final calls were to take on cargo for the West Coast. Some was loaded at Mobile, Alabama; we again took on crushed oyster shells at Gulfport; and the load was completed with rice at Beaumont, Texas. All of the cargo was destined for Vancouver, B.C.

Canada was already in the war at this time, and the wartime activities were apparent in the shipping industry. The Griffiths Company no longer had ships in the boneyard, all five of its ships sailing full time. Passing through the Panama Canal we shared the locks with the *James Griffiths*, also westward bound from the East Coast.

Coming down the East Coast the chief engineer had been quite sick. Jake Jacobi was a fixture with the company and *El Cedro*, having sailed in that position on the ship for many, many years. I would come down off watch on the bridge to find him out on deck, sitting on a bench. He said he just couldn't breathe in his room. We put him in the hospital ashore at Cristobal before starting through the Panama Canal, where the doctors told us he was suffering from dropsy. A radio message came a couple of days later, reporting that Jake had passed away. It was a sad time for the crew because he had been well-liked by everyone.

While laying in Balboa after getting through the canal, waiting for clearance to head north up the coast, we had some time to go ashore. We were warned, though, not to go up to Panama City. They were having elections, which are taken seriously there. Naturally we went up there anyway, and sure enough got caught in the riots that go along with elections in Panama. People were fighting and shooting in the streets, police on horseback were trying to hold the crowds back, and the border was closed. We finally got through all right, but were about two hours late in getting back to the ship. The old man raised hell with us over that.

It was getting close to Christmas when we called at San Pedro to take on fuel. We figured we'd arrive in Vancouver on Christmas Eve, so I wrote Gwen to meet me there to celebrate the holiday together. But the weather didn't cooperate. The trip up the coast was one of the roughest times I'd seen, with northwesterly winds and seas that had *El Cedro* buried in green water most of the time. It was so rough at about 10:00 PM one night that I called down to the captain. "Captain," I said, "it's pretty rough up here. We're making heavy weather and the ship is laboring, so I thought you might like to come up and have a look."

"What the hell can I do about it?" he snapped at me. "I can't do anything more than you're doing." He wasn't very happy with me for waking him up. I just kept the ship on course, wondering if we'd make it through the night. We did, but the storm delayed us a whole day getting in to Vancouver.

Meanwhile, Gwen had gone up to Vancouver as I'd suggested. It was hard for her to get any information about our delayed arrival because that kind of news was suppressed during wartime. But she was pretty snoopy and found ways to get the information. When we finally arrived there was a message from her waiting for me. We got together and went up to the Georgian Hotel.

The radioman from the ship had been married just before we left for the East Coast, and his bride had also come up to Vancouver to meet him. Gwen and I arrived at the hotel the same time as Sparks and his wife. The clerk told us they had one room with a double bed and one room with two singles. "I'll toss you for the room with the double bed," I told Sparks. The newlyweds weren't too happy when I won the toss.

After Christmas we made a couple of more trips south with *El Cedro*, calling at Astoria to load sacked grain for Stockton and then taking lumber down to San Pedro. In early February I got a call from the company while we were in Seattle, asking me to go down to San Pedro for a job as second mate on the *Stanley Griffiths*. Captain Ericsson objected, suggesting that his second mate take over the job on the *Stanley Griffiths* and that I stay with him as second mate. The master of the *Stanley Griffiths*, Captain Julius Johansson, had a reputation as a skipper who was hard on second mates, so I also objected. But the company insisted that I take that position.

I soon found myself on a train, bound for San Pedro and that new job. Riding through the beautiful Northwest countryside on the evening of departure from Seattle, I was going through the papers in my briefcase and found a poem I'd written about Jake Jacobi while sailing on *El Cedro*. Men at sea fight boredom in many ways. One I had picked up was writing poems about my shipmates, especially those who were real characters like Jake. My poems will never compete with William Shakespeare, but through the years they have helped me remember old friends, as I remembered that night on the train to a new seafaring job. Here are just a few of the many stanzas:

Endings and Beginnings

If you never heard of El Cedro, Jake,
    I think it's time you did.
This bird's known 'round the world,
    Started to sea when he was a kid.
So just be patient for a bit,
    While I try to analyze
The guy who's got a monopoly
    on the world's best alibis.
Now Jake and I are the best of pals,
    I'm doing this in fun.
I've known him for many of years
    and know the things he's done.
To try to pin him down to facts,
    I tell you it's no use,
Because that man has magic
    when he needs a good excuse.
When it comes to sleeping
    I'll tell you he's a bear.
It makes no difference standing up,
    laying down or sitting in a chair.
And when it comes to snoozing
    you couldn't jar him loose.
With twenty sticks of dynamite
    or a thousand volts of juice.
One day when napping,
    snoring to his heart's desire,
He lost a cigarette from out his hand
    and caught his bunk on fire.
Jake woke up with a startling fright
    and this is what he told:
"Some chump a passing by
    threw that butt in my port hole."
You may think that it was tough for Jake
    to lose his bed.
You're wrong my friend, for here again
    the old boy used his head.
So now he took advantage of
    this self-inflicted wrong,
Straight up to the master went
    and muttered him this song.

Sea Travels

"I've lost my donkey breakfast
 'count of some poor careless bloke,
Who threw a butt in my open port,
 sent my mattress up in smoke."
So sadly did he say these words:
 "I'm not to blame,
It's that reckless no-good clown
 who set my nest aflame."
Now the skipper's not so gullible
 as to take Jake at his word.
He looks at him with doubtful eye
 and says "That's the best I've heard.
A new sleeping pad I know you need,
 I'll not deny you that.
Next time put out your cigarette
 when you take your daily nap."

It's tough, losing a friend. It wouldn't be long before I'd be losing another from that same crew on *El Cedro*.

CHAPTER THIRTEEN
Sailing into Wartime

THE *STANLEY GRIFFITHS*

THE *STANLEY GRIFFITHS* was loaded, ready to sail for Panama when I arrived in San Pedro. Captain Johansson welcomed me when I went aboard. "How are you? Glad to see you!" He seemed so friendly that I wondered about his reputation of being hard on second mates. In fact, we got along well together for all the time I was to serve aboard his ship.

We proceeded on an uneventful trip to Panama where, among other items, we discharged creoscte from one of the holds. The hold was then steam-cleaned and filled with coffee beans for the return trip to Puget Sound, some general cargo completing the load. There was hell to pay when we got back to Seattle and discharged that cargo. The smell of creosote was all through the coffee beans. Between getting that straightened out with the cargo owners, and loading the ship with lumber and pilings at Tacoma, we spent enough time in port that I was able to be home with Gwen for quite a few nights.

It was May 2, 1941, when we were finally ready to sail again. First, though, the ship inspectors made us take some of the pilings off the big deck load we'd taken on at Tacoma. They felt the ship was getting cranky in her stability with such a high load.

From Tacoma we sailed direct to Balboa with no stops, shaping our course well off the coast. This gave me a chance to get back into the kind of navigation I enjoyed so much. In those days every mate navigated and put his position on the chart, with the captain checking the results. As second mate, I was the officer responsible for navigation, running the positions, calculating the day's run and other work involved with determining the ship's progress along the course laid out.

Off the coast of California the weather turned nice and balmy. It's hard to describe now what a pleasant experience it was to sail along into tropical seas on a good ship with a fine bunch of shipmates. The chief mate got the deck crew out to paint the ship and put her into nice shape during that long voyage to Panama, just under three weeks from Tacoma to entering port at Balboa.

It took more than a week to discharge all that cargo in Balboa. On June 1st we left for Puerta Chicana, Peru, to take on a full load of sugar. With no docking facilities there, all that sugar had to be brought out to our anchored ship for loading. It was June 10th when we departed Puerta Chicana bound for Texas City, Texas, via the Panama Canal. The war news from Europe and the Far East was getting worse and worse.

I had sent Gwen some letters from Peru, hoping she might meet the ship at Texas City when we called there, and again mailed some letters from the Canal Zone. There was no mail waiting for me at Texas City, so I was pretty worried about her. She hadn't been feeling well for some time. I didn't know if she'd received my letters, was too sick to make the trip, or what was going on there at home.

On a Sunday morning I hung around the ship, still waiting for news that didn't come. Finally I decided to go up into town. As I was going down the gangway a fellow on the dock asked if Mr. Christensen, the second mate, was aboard. "That's me," I told him. "Well," he said, "I just got a message that your wife is in town and she's on her way down to the ship." I was one happy fellow!

Sailing into Wartime

After she arrived at the ship I asked the chief mate if I could get some time off. He said it was all right with him, but I'd better check with the captain. "Sure, that's fine," Captain Johansson said when I asked him, "Bring her up here so I can meet her." I warned Gwen to be careful of the old boy before we went up to his cabin, as he had a reputation with women. But he was very nice to both of us and gave me the time off, saying they'd let us know when the ship was ready to sail again.

We went to Galveston, not far from Texas City, checking into the Jean Lafitte Hotel and spending several days together there. Those wonderful days ended in a sad farewell. It was hard to say goodbye with Gwen still not feeling well, the ship assigned to the East Coast, and war coming.

From Texas City we sailed for Carteret, New Jersey, with a load of potash. I was feeling depressed. Besides Gwen not feeling well, her father was quite sick in Phoenix. While we were sailing from Texas she was on her way to *Phoenix* by bus, planning to visit her dad there before going on to Seattle. I wished I didn't have to be away so long, so far from home, but sailing was my career and I didn't want to take a chance of skipping around from company to company when I had such a good thing going with the Griffiths people.

The *Stanley Griffiths* was back into the East Coast shuttle, on runs much like we had done with *El Cedro*. After discharging the potash in Carteret we went to Norfolk, Virginia, to take on pilings for Guantanamo Bay in Cuba. Next we went around to the other side of the island of Cuba to Antilla, loading sugar there for the short hop to New Orleans.

While in Carteret we had shipped some new sailors for the crew. One of them was a German fellow who served on my watch, a pleasant young sailor who was all obedience with his "Yes Sirs" and "No Sirs." Before entering the Mississippi River to run up to New Orleans we picked up a pilot. The channel is narrow at some spots along the river, requiring a good helmsman. I was surprised when the pilot gave an order to bring the wheel one way and this sailor brought it the opposite way. The pilot corrected him immediately. When the German sailor did the same thing again, the pilot said "Get this man off the wheel. I don't want him here anymore."

I told the skipper what had happened. When we docked at New Orleans there were FBI agents waiting for the German sailor. They interviewed him and then put him under arrest on the ship, with a 24-hour guard by federal marshals. During the two days they held him there I talked with the marshals, learning that this sailor was a Nazi who had been watched by the FBI for some time. "How bad is he?" I asked. "Real bad," was the reply. "What would you do if he ran down the gangway?" "I'd shoot him," the marshal told me. The old man was pretty unhappy about all this, especially when he was told that the company would have to pay for the guards. We were all relieved when the sailor was taken off the ship and we got under way again. The incident with the German sailor was just another of many signs that we were sailing into a time of war.

Our next port of call was Tampa, Florida, to load phosphate rock at a quarry right alongside the bay. We had a chance during loading there to go up into town to hoist a few. It was dark when we returned to the ship, having to walk back. Wearing tropical whites, walking along through that balmy weather, I was feeling pretty happy except that walking on the road was tough. I saw a path a few feet from the road. "I'm going to walk along that path," I told the fellows with me. I moved over onto what I thought was a path, but turned out to be a ditch full of oil. My nice white pants and shoes were black and dripping from oil, which my mates thought was a great joke. I never again tried walking on a dark path.

The phosphate rock loaded in Tampa was bound for a fertilizer plant in Boston. Sailing through the Florida Straits we got into the Gulf Stream current in nice weather on the way north, adding about three knots to our speed by staying in the middle of the stream. Fog set in as we approached Boston off Cape Cod. We steamed slowly along for a couple of days and nights, blowing the whistle every two minutes, keeping a bow lookout, trying to avoid the many trawlers and other fishermen out on the cape. Eventually we arrived at the Mystic Coal Company dock in Boston without mishap.

The phosphate rock had been loaded by chutes in Tampa, which doesn't take much time. Discharging the rock took longer, using dockside cranes with clamshell buckets. That gave us some time to get into Boston to see the historical sites and to enjoy a good lobster feed.

Instead of going around Cape Cod again when we left Boston we went down through the Cape Cod Canal into Long Island Sound, bound again for Norfolk. It was late August when we took on another load of piling in Norfolk for Guantanamo Bay. After discharging the piling at Guantanamo this time we went to the small port of Caibarien on the north side of Cuba, again to load sugar.

We were anchored out in the port, and I was working on the bridge in preparation for sailing, when a launch came out from shore with a message for me. It was a wire from Gwen that had taken several days to arrive, telling me that my mother had died. I felt very bad. I hadn't seen my mother in quite a long time, though I had written to her often. My last letter had been written on her birthday, June 30th, when I sent her another of my poems. My emotions got the best of me when I thought of the last stanza of that poem:

> *May your burdens now be a lighter load,*
> *Your remaining path a smoother road.*

Sailing into Wartime

You've done your duty as God wished you to,
May many more birthdays come to you.
As I look back on the job you've done
I'm mighty proud, that I'm your son.

There was a nice young AB on watch with me, a fellow named Pearson, who was a pre-med student at the University of Oregon, sailing now to earn money so he could stay on in school. After we got under way he said "Mr. Christensen, after you got that wire I could see that it was bad news." I told him that my mother had passed away. He said he was sorry, then asked me what my relationship with my mother had been. Good, I told him. "Well," he said, "that's very important. When something like this happens and you look your life over it's good to know that you had that kind of relationship with the one you lost." I've always remembered how that young sailor reassured me at that sad time.

Our load of sugar was bound for Weehawken, New Jersey, right across from New York City. The course took us up through the Bahamas on a nice moonlit night. That bad news heightens my memory of the time. I remember Captain Johansson coming up on the bridge and saying "Mister, take a sight on the moon." There was really no need to take a sight. In the bright moonlight we could take bearings on land. But I took the sight and worked out a position. The captain felt good about that. He knew I liked to navigate and thought I could do anything that had to do with navigation.

Our orders after delivering the sugar to Weehawken were to set us off on a new adventure. We were to go to Norfolk again, but instead of loading pilings for Cuba this time we were to load coal for Rio de Janeiro. It's a long haul from Norfolk to Rio, a change from the shorter trips we'd been making around the East Coast. It was overcast and cloudy when we finished loading and departed Norfolk on September 18, 1941, setting a course that would take us outside the Bahamas on the way south.

The next day it started to blow, coming to a gale force with seas starting to build up. We weren't making much headway, pitching into those seas. The next day was even worse. With that heavy load of coal we were well down into the water, with seas breaking over the bow regularly. Built of riveted steel plates the *Stanley Griffiths* was a sturdy ship, able to take a lot of weather. But that wasn't much comfort when the winds built up to hurricane force on the night of September 20th. Now we were making no headway at all, wallowing in those terrible seas and constantly awash. The masts were about all that could be seen of the ship from the bridge. The seas were breaking so high over her that some were washing into the engine room through the fiddley, the entrance to the engine room from the upper deck. Even the dining saloon and cabins were awash, water sloshing to and fro. When one of those heavy seas hit us, it felt like an earthquake. All the spars and awnings were quickly washed overboard. Later, sometime during the night, we took a wave so huge that it tore the emergency steering wheels and gear right out of the after deck and overboard.

None of the crew showed any fear during the storm. We had a lot of faith in that ship. It was a relief, though, when the weather finally started moderating. That was on the next day, September 21st, after we had been hove to for thirty hours. Our next task was to assess the damage. We were lucky that the hatch tarpaulins had stayed on and the hatches were secure. Other than the loss of the awnings and steering gear, and the mess we had on our hands with all the water in the cabins, we were in surprisingly good shape for the beating we'd taken.

This area where we hit the storm is today called "The Bermuda Triangle," and lots of mystery is attached to the loss of so many ships there. There's no mystery to it. It's simply an area that gets some terrific storms, storms so strong that some boats and ships just can't weather through them.

It was sad to learn later that the *Libby Main*, the ship I'd made one trip aboard to Alaska, went down near us in this same hurricane. A wooden ship, she had been loaded with railway iron and just couldn't survive those winds and seas. Worse yet, our old friend Two-Step Harry was chief mate aboard her at the time and was lost along with all but two of the crew. Three of the crew managed to get off the ship and onto the top of the ship's cabin, which had broken loose and was floating. One was Andy Abramenkoff, a mate I was to know in later days. The ship's cook and a seaman were with Andy on the cabin top, but the cook was swept off and lost before the other two were picked up after a few days. The two survivors were in terrible shape when they were rescued, spending some six months in the hospital before they were able to work again.

One thing about the sailor is that he always forgets about the last storm he's gone through. Going down to the union hall for the next job, his mind is on the pleasant voyages through good weather, not the storms he's experienced. As we steamed away day after day, getting further south into nice tropical weather, the crew was cheerful and there was no talk of the beating we'd taken in the hurricane. I was especially happy because sailing so far south brought new stars for taking sights, getting me deeper and deeper into the celestial navigation that I liked so much. It was a leisurely, pleasant trip to Rio de Janeiro, twenty-five days from port to port.

We laid at anchor in Rio, discharging the coal into an old ship tied alongside us, then moving to a dock to load ore for the return trip. Our second engineer got sick one evening while I was on watch, moaning and groaning and in such bad shape that I called a doctor. The doctor gave the engineer a shot to calm him down, then told me we'd better get him to a hospital. Since there were several railway tracks between the dock and the nearest road, there was no choice but to carry the engineer on my back up to the road to hail a cab and take him to the hospital. After getting him registered and into a bed there, I shopped around town for a little while before returning to the ship. I was surprised to find the sick engineer already back at the ship. "I'd rather die on the ship," he told me, "than spend any time in that hospital." He didn't die, and he never told me what had frightened him so much in the hospital.

I got myself in serious trouble while we were in Rio. A few nights before the ship departed, a bunch of us had gone up to the Copa Cabana nightclub to celebrate. We celebrated so much that I was pretty loaded by the time we returned to the ship in the early morning hours. I peeled off my clothes, getting ready for bed, then left my cabin for a minute to talk to the mate, wearing only my skivvies. When I got back to the cabin I thought my clothes were missing. I was certain I'd put them on my bunk, but they weren't there.

We had all become touchy about theft. Lying alongside the old ship taking our coal, there had been lots of thievery by the work crew trimming the coal as it was loaded into that ship. I went out on deck to raise hell about my clothes, telling the customs officer guarding the gangway that I was going to go over to the old ship to get them back. "You're going to go with me," I told him. He started arguing with me, I kept telling him he had to come with me, and then he drew his revolver. I still don't know why but I disarmed him, taking the revolver myself.

That was the wrong thing to do. He started blowing his whistle and waving a flashlight towards shore. Pretty soon a launch came out and smoothed things over. But in the morning the chief of the customs office and other officials came out to raise a stink about it. They were going to arrest me. I was pretty embarrassed when they held a hearing in the captain's quarters, as by then I'd discovered that my clothes hadn't been stolen after all. I've always been neat about hanging up my clothes when I go to bed, and even in my loaded condition that night I'd done just that before I went to see the mate.

Captain Johansson proved to be the best defense lawyer a man could have. He argued that they shouldn't take me ashore. "That young fellow," he said, "has been with me a long time. He's a clean-cut young man who never goes out carousing. He always comes back to the ship sober, but this time he probably did have something to drink. It's no wonder it affected him, he's just not used to drinking. You should excuse him for this one time in his life when he drank."

The captain's appeal worked. I was not allowed to go ashore again, which didn't make much difference since we were sailing in a couple of days. It was a lucky break for me that Captain Johansson came to my defense. I could have been in real serious difficulties if he hadn't bailed me out with that speech.

Sea Travels

On October 26, 1941, we sailed from Rio bound for Trinidad. It was a beautiful trip up the coast, spoiled a bit by the captain having appointed me as medical officer. One of our firemen had hurt his leg while he was ashore in Rio, then hurt it again on the gangway when we were departing. He was a rotten guy to start with, and now his leg was rotten. He had an awful sore that I had to clean out with a solution, using a syringe to get right down in the wound. The leg was so rotten that it looked like a sprinkling can when I used the syringe. This fellow was so ornery that he never showed any gratitude for the help he was getting, even to his mates from the engine room gang who were washing his clothes for him. When one of them brought the clean clothes up to the fireman, he started complaining that they weren't folded properly. That was the end of any help he got from them.

When we got to Trinidad, where we refueled, a doctor came aboard to look at the fireman. He washed the sore out some more and gave me a different solution to use on the leg. I'd hoped that they'd take him ashore in the hospital there, but no such luck. He stayed with us all the way to Baltimore, where we did get him into a hospital. We never learned what happened to him, but I imagine he lost the leg.

The signs of war were closer to us all the time on this trip. German submarines were already patrolling these waters, looking for ships of those nations such as Great Britain and Canada with which they were already at war. We had our flag painted large on the side of the *Stanley Griffiths*, and weren't too worried about danger from the submarines. Of more danger to us were floating mines, some of them from way back to World War I. Before arriving in Trinidad we spotted one of those, about 200 yards off the ship, reporting its location to the authorities. And then while I was on watch still closer to Trinidad, I spotted a sunken vessel on Noble Shoal. The top part of the bridge, the masts and the funnel were all that could be seen of her. When we asked about the wreck in Trinidad, they told us that she was a Canadian ship that had come in to refuel a few nights before and was sunk by a German submarine on departure. They still had no word at that time on the fate of the crew.

Our cargo for this trip north to Baltimore was manganese ore, which we discharged at Sparrow Point on arrival there. It was the 20th of November, very cold as I remember. Sparrow Point is quite a ways from the city, but the chief mate and I decided to go up into town to get haircuts, much overdue after all that time at sea. Riding along in the cab we spotted a pub. It looked like a good place to stop. When we went into the pub we saw they had hot brandy drinks boiling away over a flame, just the thing for a cold day.

After a couple of those we went to get our haircuts, then decided to try a shore dinner for a change. We found a restaurant with some entertainment, where we had dinner and a few highballs while we listened to the singers, a good time after weeks at sea. As we started out of the place a big fellow grabbed us and said "Get in there," showing us to a paddy wagon backed up to the door of the restaurant. "I'm not getting in there," I told him. He was a policeman, big and tough. "The hell you aren't," he said. When I asked him why I should get in, he told me never to mind why, just get in. So we climbed into the paddy wagon, along with all the other guests from the restaurant, and they took us to the Baltimore city jail. When we got to the jail there were two or three hundred other people who had been rounded up at various places around the city. We were fingerprinted, run through a line, and then had to pay $2.50 bail to get out of jail.

As I was paying my bail the clerk asked "Where are you from?" "Seattle," I told him. "That's a hell of a long way to come to get caught in this racket," he said. I read in the paper the next day that almost five hundred people had been picked up in the raids, which had been run because the jail needed money. We could have gone down the next day to get our $2.50 back, but never did. I don't know why, but somehow I couldn't keep out of trouble on this trip, even when I was innocent.

When discharging was finished we went into a drydock at Baltimore for the required annual inspection. The inspection, painting the ship's bottom, and some minor repairs took a few days. The shipyard was jammed with work, every imaginable kind of vessel being hauled out of the boneyard to be made seaworthy, many of them destined to be turned over to the British for wartime use.

Our next orders were for Charleston, South Carolina, to load all kinds of things for the Philippines. Tons of mats for temporary landing strips, Army trucks, lumber and other cargo useful in wartime was put aboard. We took on stores for the long voyage and were planning to take on fuel at the Panama Canal Zone. The *Stanley Griffiths* departed Charleston early on the morning of December 7, 1941. I was on watch on the bridge and we had just cleared the harbor when the radioman informed me that Pearl Harbor had been bombed that morning. We were having trouble believing the news when orders came for us to turn around and put back into Charleston Harbor, returning to the Army pier where we had taken on the cargo.

We lay there for two days, wondering what would happen next. What happened was that painters came down to the ship with spray guns. Starting at the tops of the masts, they painted the ship gray all over. Even the lumber and trucks on deck were gray by the time they finished, only a few hours after the work was started.

Another few days passed while we lay at the Army pier, waiting for a decision about our voyage. Finally, they took all the cargo out of her. Next they fitted her with guns, antiaircraft weapons forward and a .375 cannon aft. Meanwhile I had been trying to get through on the telephone to Gwen, with no luck because the circuits were jammed. When I did get through to her at last I found her sick and scared. Seattle was in the middle of a panic, with fears that the Japanese would attack there next. The city was blacked out at night and all kinds of crazy rumors were passing around. With Gwen facing a major operation, I offered to come home to be with her during that time. She left that decision up to me.

I had a commission as an ensign in the U.S. Naval Reserve. Still in a quandary what to do, I went down to the Naval Recruiting Office in Charleston along with our second engineer, to volunteer for active duty. We went to the office two or three times with no decision being made on our situation. On our last visit we talked with a Navy captain, who asked about our licenses in the merchant marine and our experience. When we told him, he said "You fellows stay right where you are. Men like you are needed in the merchant service. We can make ensigns in 60 days." We took his advice and went back to the ship.

It was late December, and they were completing installation of the guns and quarters for the gun crew, when I talked with Gwen again by telephone. She was sicker than ever, and had scheduled herself for surgery. With that I decided to go home to be with Gwen during the surgery, and while home to study and sit for my chief mate's license. I felt bad about leaving the ship with the war just started, worried that the rest of the crew would think it was a cop out. Captain Johansson was very nice about it. He said he really hated to see me leave at such a time, with other new crew coming on besides my replacement, but he understood the situation.

I booked myself on a train for Chicago. Just as with the ships, the railways were hauling out all their unused equipment because of the wartime needs. We boarded the train on a siding about midnight. The car I was in was an old one with hard benches along the sides and a potbelly stove for heat. We had to change trains in Atlanta, Georgia, when I was given a chance to change to the Dixie Flagler, a faster train. I jumped at the chance, catching the Milwaukee Railroad's old Empire Builder in Chicago and arriving home not long after New Year's Day.

Gwen had her operation as scheduled, spending about ten days in the hospital before coming home to recuperate. Meanwhile I enrolled myself into Captain Kildahl's Navigation School, studying for my chief mate's license. Besides taking care of Gwen and studying for the license, I sometimes went down to ships loading at the waterfront to take a job as night mate. Everything was running 'round the clock in those wartime days, including loading and offloading of the ships. When mates supervising cargo handling wanted a night off, a licensed mate would be hired to handle their duties.

Sailing into Wartime

I sat for my chief mate's license and got the ticket about mid-March. Gwen was recovering well by then and back at work, so I decided it was time to return to sea. She didn't want me to go. We had been separated so long already, and the merchant marine was so dangerous in wartime, that the prospect of my going back to sea frightened her. But I felt my profession was at sea and that they needed men like me in wartime. I called and told the Griffiths Company that I was ready to take a job again.

It was sometime in April that they called me. The company was acting as agent for the War Shipping Administration in operating a new ship being completed in an Oregon shipyard, a Liberty-class vessel named the *Jonathan Edwards*. The job offered was as second mate. With my brand new chief mate's license I was a little reluctant to ship as second mate again, but the other mates had already been hired. I took the job.

CHAPTER FOURTEEN

From Mate to Master

CONSTRUCTION was still being finished on the *Jonathan Edwards* when I reported to the Kaiser Shipyard at Portland for my new position aboard her. The crew lived ashore for awhile until she was completed, visiting the shipyard daily to get acquainted with our new ship. The chief mate was an old captain from the Bol Line on the East Coast, John Johansson, and the skipper was Captain Eckert, a man who had retired from the sea quite a few years earlier and had his own insurance agency in Puyallup, Washington. The third mate was a young man with a brand new license, Eric Forbes, a nice lad.

This was my first experience on a Liberty-class ship. These vessels were constructed of welded steel plates, making them very limber in heavy seas when light. When they were first put into use there was some question about their reliability. Some problems did develop, with cracked plates and one or two ships breaking up in very heavy storms, but in general they proved to be very reliable and the 2,700 ships of this class were a great addition to the merchant marine fleet so badly needed in wartime. Many served well in the U.S. and foreign merchant fleets for years after World War II.

After a shakedown cruise we headed for San Francisco empty of cargo. The ship acted in seas just like I'd heard, with the decks bouncing up and down every time we hit a big wave. Other than that everything seemed in good order and I was happy with both ship and crew, until I had a run-in with Captain Eckert. I had taken a noon sight and was running up our position off the coast, one of my duties as navigation officer, a duty assigned by tradition to the second mate. I had just finished this work in the chart room when Captain Eckert came in. "Here's the noon position, Captain," I told him, showing him the paper on the chart table. He brushed it off the table rudely, looked me straight in the eye and said "I don't trust any man. I do my own navigating."

That took me back. "Well, Captain," I said, "it's your privilege to do what you want to do. I've been navigating for a long time and I sailed with a captain who was a very meticulous man. He trusted me with the navigation on his ship and I have a lot of confidence in myself. If you don't trust me I'll just keep taking the positions, but I'll keep them to myself." And that's just the way it was for all the time I sailed on this ship. He never looked at my positions, or those the other mates ran up, doing all the navigation himself. We got along well otherwise, though after that incident I can't say that he was a favorite among the many skippers I had known through the years.

In the Bay area, we loaded lumber at San Francisco and then went to Alameda to put two big airplanes on deck. There was a terrible accident during the loading of those planes. After getting the aircraft aboard and lashed down, we were putting some planks ashore that had been used in the process. As second mate I was supervising the operation on the after deck. The supercargo had just left the ship and a man was walking on the dock when a plank fell from the sling being used to transfer them to the dock, hitting him edgewise on top of the head and killing him instantly. I thought that wasn't a very good omen for heading out to sea with a new ship.

We assembled in convoy for the voyage to the Hawaiian Islands, my first experience at handling a ship in convoy. There were about forty or fifty ships, with lots of protection by destroyers and other naval vessels. It was an uneventful trip. Sometimes we'd see navy cruisers under full steam heading westward, to what destination or battle we never knew. All the ships were blacked out, so we had to keep our position by watching the ship abreast of us. The course and orders had been pre-determined before leaving San Francisco Bay, since there was a complete radio blackout. Our convoy zigzagged along that course, each change signaled by whistle from the lead ship in the convoy.

We found Pearl Harbor still in an awful mess from the raids on December 7th. The *West Virginia* had just been raised from the bottom and moved into drydock. The towers of the *Arizona* were above water, and her hulk could be seen on the bottom through the clear water as we passed by. One of the engineers was standing with me at the rail as we watched. "I have a brother down there, asleep," he said.

After discharging the deckload of aircraft and other cargo at Pearl Harbor we went to Kahului on the island of Maui, to load sugar. The chief mate, Mister Johansson, was one of those fellows who likes a drink now and then. The first afternoon there he and I went shopping, then found a beer garden where we had a couple of scotch and sodas. The next afternoon he wanted to do the same thing. "I'm a married man," I told him, "I can't afford to go out drinking every day." "Oh, don't worry about the money," he told me, "the Chinese running my prune ranch back in California are making plenty of that for me." He was another officer who had retired, then returned to the sea when war came along. I objected but he insisted, so we went back again that afternoon. Mister Johansson wasn't a bashful fellow. He asked the Kahului Railroad for a car to go sightseeing. They provided a limousine with a driver, and we had a chance during those days of loading sugar to drive all over and see the sights of the island. Meanwhile, young Mister Forbes was left at the ship to oversee the loading.

Several days passed like that before loading was completed and we headed for the Panama Canal, bound for New York. In an earlier chapter I described that lonely voyage of eighteen days, traveling alone rather than in convoy, and the trip up the East Coast in convoy when the group was attacked by submarines. I didn't mention, though, our approach to Panama.

There was a Navy patrol boat that showed up astern of us as we neared the Canal Zone at daybreak. Mister Forbes was on watch, with the captain conning the vessel. "Captain," Mister Forbes said, "that patrol boat is signaling us."

"Mister Forbes, you're supposed to look forward, not aft. The stern will take care of itself." The captain seemed annoyed that this young officer would disturb him like that.

"But he's signaling for you to stop." Mister Forbes could read some Morse code and was following the signals from the patrol boat.

Now the captain was beginning to get irate. "Mister Forbes, I told you to look ahead, not aft. We'll just keep going here and you don't pay any attention to him. I know what I'm doing."

This young third mate didn't like that very well. Being a gutsy fellow, he called the signalman from our Navy gunnery crew up to the bridge and asked him to read the signals from the patrol boat. The signalman read the message: "Stop immediately, you're in a mine field!"

When Mister Forbes passed that on to the skipper the ship was stopped immediately. The patrol boat came alongside and raised hell with the captain for not obeying the signals right away, then instructed us to follow him out of the field so we wouldn't hit any mines. In all fairness to the captain, he was following the normal course into the Zone and had received no warnings about the mine field before departing Kahului. It was an example of the kind of foul-ups that can happen in the early months of a war, but it also shows that the watch on a bridge should look aft once in awhile.

In my earlier description of this voyage I mentioned the convoy up the East

Coast, fifty ships protected by destroyers, a dirigible and bombers escorting us overhead, and the submarine attack hitting our group after we left Key West. We lost the two last ships in our convoy in that attack. There were many, many merchant ships lost off that coast in the early war years. The shelf out from the shoreline is not very deep, and we saw the masts and funnels of many of those sunken ships sticking up out of the water as we proceeded up the coast. I don't remember how many were lost in that area, but I have an old newspaper article showing that in a six-month period a year later there were thirty-three merchantmen sunk by German submarines in the Gulf of Mexico alone.

One nice day while we were steaming up the coast in the convoy, Mister Johansson was walking aft on deck when he passed one of the firemen, an Irish fellow who was a real wise guy. The fireman had been needling the chief mate all during the voyage, mimicking Mister Johansson's Norwegian accent. On this occasion the fireman bumped into him and mimicked his accent again. Mister Johansson ignored him, but then when walking forward again the fireman made another remark. Mister Johansson didn't take it this time, giving the fireman a blow that knocked him out.

There were strict rules against officers laying their hands on members of the deck or engine crew. A complaint was filed right away with the captain, who ordered me to sign the log. Captain Eckert had logged the chief mate for this infraction of the rules, showing the charge in the log and fining him two hundred dollars.

"He's my friend," I told the skipper, "can't you get someone else to sign the log?"

"Wait a minute." said the captain, "I'm doing this for his own protection. I'm logging a fine of two-hundred dollars because there's no court in the United States that can convict a man twice for the same crime. I'm doing it now so he can't be hauled up by the board and have his license revoked."

I didn't feel very good about it but I went ahead and signed the log, hoping the captain was right and wasn't just using this as an excuse for punishing Mister Johansson.

When we finally got safely to New York we docked on the Brooklyn side, where there was frantic activity in loading all kinds of ships. Many of them were bound for Russia on the run to Murmansk, carrying assistance to a country that was then our ally. After discharging the cargo we laid there for a couple of weeks waiting for orders. It was decided that we would join one of those convoys to Murmansk. Work was under way to reinforce the forepeak of the ship with concrete, protection against the ice on the run, when the manager of the Griffiths Company visited New York.

Mister McDowell and I had a long conversation one day soon after he arrived. After we'd been talking for awhile he asked me if I would like to ship as chief mate on the *James Griffiths*, with my old friend Julius Johansson as captain. "That would suit me just fine," I told him. I liked both that ship and the skipper. Arrangements were made, and I signed off the *Jonathan Edwards* on August 28, 1942. When I was paid off, Captain Eckert gave me a very nice letter of recommendation. If he wasn't the most pleasant captain I'd served for, at least he was fair.

Transportation was so crowded during the war that it was a couple of days before I could get passage on a train home. I was happy to be going home, but it happened to be a sad time for the Christensen family. My youngest sister, Ellen, who was married and had a two-year old son, was suffering from leukemia. My trip coincided with her being in a Portland hospital, where they were trying some heroic efforts to save her life. Not much was known about the treatment of leukemia at that time. They had removed her spleen, and needed a great deal of blood for transfusions. The train I was on passed through Portland, so I stopped off to visit her and to donate some blood. She was transferred to a hospital in Port Townsend not long after that, where she passed away in mid-September.

As if that weren't enough, my dad was involved in a serious collision in Alaska with the mail boat about the same time. He was returning from his usual late summer in Wrangell and had gone to sleep soon after departure, leaving one of his crew to navigate. They were just south of Lincoln Rock when the *Hannah C.* plowed into the port side of the mail boat, holing the other boat and badly damaging her own bow. Fortunately, the mail boat didn't sink and Dad was able to escort it safely into port. He took the *Hannah C.* to Ketchikan for repairs, where he learned of Ellen's death and flew home for the funeral.

Dad's boat was clearly at fault in the accident. Since he was still under bareboat charter to the cannery the claim was handled by the cannery's insurance. Paul, who was home at the time, flew up to Ketchikan when repairs were finished and brought the *Hannah C.* back home. With my mother's death the previous year, Ellen's death, and the fact that Paul and I could no longer crew for him, Dad was pretty discouraged. He sold the boat he had built and run for so long, going to work for the Fish and Wildlife Service as a skipper and mate on a number of their boats working in Alaska.

The reason for Paul's being home was still another accident. He had been working as chief mate on the *Elna*, a steam schooner hauling freight to Alaska. They had been running in poor visibility down Chichagof Strait when someone applied compass deviation the wrong way. The ship piled up on Chichagof Island, breaking up and sinking. Luckily there was no loss of life.

Despite all these sad events it was good to be home again. Gwen and I were especially happy to spend some time together. The *James Griffiths* had just departed for Alaska when I arrived in Seattle in early September, so I had almost a month off from work before her return.

And there is always some bit of humor to lighten the burdens of life. While I was home I ran into Ole Monsass, the mate who had befriended Gwen on her trip to Alaska aboard the *Northland*. Ole and I were talking about Paul's accident with the *Elna*, and while talking about him Ole told me a story from the time when Paul was a winch driver aboard the *North Sea*, a passenger steamer running to Alaska for the *Northland* Transportation Company. Ole was chief mate on the ship.

They had just left one of the Alaskan ports, and Paul had finished securing the winch gear, when he saw a lady on deck with a little terrier. "What a nice looking little dog!" Paul said, leaning over to pet it. The lady said "Yes, I'm awfully fond of it. I bought it while we were in Seattle. I wish we'd had its tail amputated, though; it's too long for the breed."

"Oh," said Paul. "We can take care of that. There's a veterinarian aboard ship."

"How nice. Do you think he could take care of that?" The dog's owner was pleased that her problem could be handled right on the ship.

"I'll take care of it." Paul picked up the dog and took it down to the butcher shop alongside the galley, laid the dog on the meat block, took a cleaver and whacked off the dog's tail. Then he pasted a bandaid on the stump and carried the dog up to the lady's stateroom.

"That didn't take very long. Done already? You must have a very good veterinarian on the ship."

"Yeah, he sure is." Paul handed the dog to the lady. "That'll be five dollars." The lady apparently never got wise to the fact that the "veterinarian" on the *North Sea* was just Paul with a meat cleaver.

It wasn't until mid-October that the *James Griffiths* returned to Seattle from her Alaska run. Captain Johansson was tickled to see me. He hadn't been pleased with the mate who was sailing with him. I was happy to be back aboard. I liked the captain, knew the ship well, and being on an Alaskan run I'd be home more often than during those long runs on the East Coast.

We loaded cargo at a number of ports around Puget Sound. As chief mate I had responsibility for the loading, finding myself very busy during those days. Our first port of call was to be Women's Bay on Kodiak Island. Because of the war we first went up the Inside Passage to Cape Spencer, gathering with other ships to join in a convoy across the Gulf of Alaska. Running in these convoys was a lot more difficult than the others I'd experienced.

It was early November with cold, blowing, snowing, misty weather making it hard to keep track of the other ships in the convoy. It was wise to be in convoy, though, as there had been some Japanese submarine attacks on merchant ships in these waters, with one or two sunk.

On this trip we arrived at Women's Bay without incident, discharging our military cargo there and returning to Seattle empty. We loaded more supplies around the Sound, taking on ammunition at a dump at Mukilteo, then took a three-masted schooner in tow, bound for Excursion Inlet near Juneau. The schooner was an example of the fact that every possible resource was used during the war. She was the *Charles R. Wilson*, the same cod fisheries vessel we had towed in from Cape Flattery so many years earlier when I was engineer on the *Phillis S.*, still carrying all three of her masts. The Army Engineers had chartered her for their use.

We had just finished getting all the gear secured on deck and were under way for Alaska. I was sitting in the dining saloon having a cup of coffee when one of the sailors came running in. "Mate, you better get out on deck in a hurry. The chief cook has fallen from the boat deck to the deck. He's hurt."

I found the cook with one ear nearly torn off and his head bashed in. As soon as I could notify the captain we turned around with the schooner still in tow, radioed to the Coast Guard in Seattle and ran into Elliot Bay to transfer him ashore. He didn't live very long. The sad part of this incident is that we had waited a couple of hours for the cook before we could depart, because he'd been up in town celebrating. He was a little tipsy when he came aboard, was standing on the boat deck talking to some friends in the crew and leaning against the boat deck rail, when he lost his balance and went backward over the rail and down. He was a very nice red-headed Norwegian fellow, and an excellent cook. It was sad to lose him that way. We often thought that if we'd just not waited for him he would not have died.

We got under way again, towed the schooner to Excursion Inlet, then ran to Cape Spencer to join another convoy bound for Women's Bay. Puget Sound to Women's Bay was to be a regular run for the *James Griffiths*, a tough one during winter months but a good job as far as I was concerned. It was sure a lot better than being off on the East Coast or out in the Far East.

On one of the trips back to Seattle that winter we had a young Navy seaman stow away on the ship, not discovered until we were close to Seattle. He was a kid only sixteen years old, who lied about his age to get into the Navy and then discovered he didn't like it much. The MP's met the ship and hauled him off when we docked. The poor kid was scared to death, but I doubt that they gave him much punishment considering that he was only sixteen.

On another trip we carried sixty cases of liquor for the Navy. Designated "for medicinal purposes," it was good John Dewar's White Label whiskey. While the cargo was being discharged by Navy sailors at Women's Bay there was a Marine detachment there to keep an eye on the liquor. Mixing sailors with cases of liquor belonging to someone else is asking for trouble. It was no great surprise to us when the Navy took inventory and found several cases of their whiskey missing.

That caused a real stink. The Marines were sent back down to the ship to conduct a search. They found occasional bottles of different kinds of booze stashed away by our own sailors, and even a bottle of Dewar's Scotch Whisky in the quarters of our Navy gunnery officer, but no sign of the missing cases of whiskey. Not long after that I saw one of the Navy sailors in the cargo crew fooling around at one of the life boats on the boat deck. I went over to take a look and found that some of the missing whiskey was in the boat, which the Marines had failed to check. I threatened to report him to his commanding officer, but he pleaded so much that I finally agreed to let him go with the promise that he'd never again come up to the boat deck.

The Navy wasn't ready to give up. We were ordered out to anchor in the bay, and the Marines were sent out for another search. This time they even ordered the captain to open the ship's safe. We had a relief skipper on this run, Captain Shay, because Captain Johansson was suffering from stomach problems and hospitalized in Seattle. Captain Shay refused to open the safe, telling the Marines it would take much higher authority to persuade him to open it. Once again they found nothing, and that was the end of our whiskey affair.

Captain Johansson had returned to the ship when, after several trips like this in convoy, Navy Intelligence decided that the submarine danger had passed enough that we could start running again as single ships. Following that decision we sailed direct from Puget Sound to Kodiak Island without running up the Inside Passage or sailing in convoy. These were uneventful voyages, if sailing the North Pacific in winter can be considered uneventful. There were gales, there was snow, and rain if it wasn't snowing, with visibility so bad that it was very seldom we could take sights for celestial navigation.

We lost a man overboard while on one of those runs between Cape Flattery and Kodiak Island. When I came on watch at 4:00 AM, well out to sea in the midst of a storm, I found a note in the log that the utility man in the galley was missing. The mate I was relieving was unsure whether or not the captain had been notified, so I called Captain Johansson to report the incident. The captain already knew about it. He told me he'd been called, they had searched the ship and were unable to locate the missing man. It was assumed that he had gone overboard just aft of the midship house. We were carrying a deck load of lumber and trucks, and had proper lifelines run around the load to prevent someone slipping over. It was never learned exactly what happened to this crew member, but our guess was that he slipped under those lines and overboard while throwing a bucket of slops into the sea.

We committed another man to the sea on another of those runs, under much different circumstances. The general manager of the Griffiths Company, Arthur Wuthenow, passed away during this period. His wishes were to be cremated, and that the ashes be cast upon the sea from one of the company's ships. The lot fell on the crew of the *James Griffiths*, again on the run between Cape Flattery and Kodiak Island. One forenoon after we were out on the Pacific, Captain Johannson called me and said it was time to perform the ceremony. The chief engineer was asked to join us, the three of us being those in the crew who best knew Mister Wuthenow. It was raining and stormy as we went aft with the ashes. Captain Johansson gave a little service before we scattered the ashes on the sea, an emotional moment for me. Mister Wuthenow was the man who gave me my first job as an officer with the Griffiths Company, so many years earlier.

Those voyages to Kodiak Island were more pleasant as summer approached. When we returned from one of the trips in the first week of May 1943, I had enough time in as chief mate to be able to sit for a master's license. Captain Johansson was very nice about it when I told him I wanted to take off to study for the ticket and sit for my license. He gave me a fine letter of recommendation. The mate who relieved me was Andy Abramenkoff, one of the two men saved from the *Libby Main* when she went down in that hurricane in the Caribbean.

Back to Captain Kildahl's Navigation School. I enrolled a few days after leaving the *James Griffiths*, finding that the studies went easier with each examination up the ladder due both to earlier studies and the experience at sea. It was about mid-June that I sat for my master's examination, eight hours a day for four or five days. Taking my new license proudly up to the Griffiths Company, I was congratulated by Mister McDowell. He then asked me if I had enough time in to sit for a Puget Sound Pilot's endorsement. That, he told me, was a requirement to serve as master of one of the Griffiths ships. I did have enough time to qualify, so I gathered the records to prove it and went back again to Captain Kildahl's school. More weeks of studying, and more days of examinations gave me that endorsement.

This time I got my first job as a ship's master when I went back to the Griffiths Company. The *Stanley Griffiths* had been in dry dock at the Todd Shipyard in Seattle, due to some bottom damage she'd suffered. Coming down through *Tongass* Narrows near Ketchikan in a heavy wind, empty of cargo, she'd lost steerage way and hit a reef. The captain, another Christensen who was no relation, was not held to blame under the circumstances, but had to leave the ship for some other reason. I took command in the latter part of July 1943 and took her into the Stacey Street dock for loading. Some of the company men came down when we were finished, asking if I felt confident about my new command.

Sea Travels

"I have all the confidence in the world," I told them.

Getting out of our moorage looked difficult. There were ships on either side of the docks astern of us. It was a long and narrow passage to back out of there. "You think you can get her out of here?" they asked me.

"All I need is one tug boat." We got a Red Stack tug down there and I put him on my port bow, backing out and turning around in the waterway with no problem. I found that handling a large ship was not much different from the many smaller boats I'd handled as a younger man. The main difference with a larger ship is that you can't feel the movement as you can with a smaller one, but with so much experience watching these ships being handled from the bridge I could predict what the ship would do in any circumstance.

We were bound for Anchorage. Since I didn't have a pilot's endorsement for Alaskan waters we picked up a pilot, went to the Union Oil dock at Edmonds to take on fuel, and then up through the Strait of Juan de Fuca to poke our nose out into the Pacific at Cape Flattery. I was a skipper at last!

CHAPTER FIFTEEN

The *Stanley Griffiths*

THE APPROACH TO ANCHORAGE is a difficult passage through Cook Inlet, famous for its strong tides, foul weather and shallow waters. Until we arrived at Cook Inlet, our pilot on this trip had nothing to do. He and I were both on the bridge as we entered the inlet. Getting out the charts, I said "I've never been up in Cook Inlet."

"Neither have I," he said.

Well, I thought, at least we're starting out even. By the time we got up to the point where the Kenai River enters the inlet the tide was running like a millstream and heavy fog had settled down over us. I decided to drop anchor right there in the middle of the inlet. The water is so shallow that it's possible to anchor almost anyplace there.

Our anchor held through the night against that strong tide. At daybreak we weighed anchor and headed on up the inlet. I decided to take the *Stanley Griffiths* into port myself. It was silly to have that pilot along, considering that he'd never been there himself, but it was a union rule to carry a pilot if the master didn't have a pilot endorsement for the waters.

Laying at the dock in Anchorage to discharge cargo was a challenge in itself. The tidal changes are so great that we had to breast the ship out with barges between the ship and the dock, unloading onto them. A dock crane then picked up the cargo from the barges. There was another reason for breasting the ship out this way. The water next to the dock was so shallow that we might otherwise have been sitting on bottom at low tide.

Everything went well in discharging the cargo. During the process I had a chance to visit Anchorage, the first time I'd been there in all the years of sailing to Alaska. We returned with a light ship to Seattle, where we loaded general cargo for Kodiak and Whittier. Taking the ship into Women's Bay at Kodiak was no problem. I'd been there so many times with Captain Johansson that it was familiar territory. I'd also been to Whittier many times, where we discharged the remaining cargo.

At Whittier, I had to make one of those decisions captains are paid to make. We were under charter to the Army Transportation Service at the time. The port captain at Whittier sent orders for us to take a barge in tow for delivery to Excursion Inlet in southeast Alaska. The *Stanley Griffiths* was very light, as we had not taken on any return cargo for Seattle, and we were not well-equipped for towing. Under these circumstances, I felt it was too hazardous to accept the tow. I sent a letter refusing the tow to the port captain, the first of countless decisions of this kind I would be making as a ship's master.

At the end of this second trip, I had to ship a new mate. The chief mate with me during the first two trips was a fine officer, an older Finn named Sundell. Mister Sundell had more licenses and endorsements than I held at the time, but I was made captain of the *Stanley Griffiths* over him because I was senior in the company. I was afraid he might resent me as captain, but he never did and we had an excellent relationship. Now he was leaving to be master on *El Cedro*.

My new chief mate was Blackie Howes. We had been together on *El Cedro* when he was second and I was third mate. After Blackie came aboard we proceeded to Vancouver, British Columbia, to take on a full load of wheat bound for San Pedro. It was a heavier load than usual. During the war years, we were allowed to carry a certain percentage of tonnage over that permitted for each ship in peacetime. In the case of the *Stanley Griffiths*, that brought us down so the water was six inches over the Plimsoll line, the mark on the side of the ship which shows the usual limit of loading.

When we got outside Cape Flattery we hit one heck of a blow. Being so deep in the water with all that wheat we were awash most of the time. One of the life rafts was lost in the storm. When we got to San Pedro and started unloading, it was discovered that an oil line had broken during the storm. Oil had damaged much of the cargo in number two hold. The steamboat inspectors came down to the ship to inspect the damage, asking that we pull up the flooring over the deep tanks to locate the oil leak. It wasn't very serious, so we had that repaired and were soon under way again with a cargo of ammunition from Port Huenene, California, bound for Alaska. We topped off that load of ammunition with some more general cargo from Puget Sound, then it was back to Alaska, calling again at Kodiak and Whittier.

THE *STANLEY GRIFFITHS*

These Alaska runs, with an occasional trip to California, were routine up to late January, 1944, when I received orders to load a cargo in Puget Sound bound for Honolulu. In addition to general cargo in the holds, which included several-hundred tons of dynamite, we had lumber on deck, two torpedo boats loaded athwartships, and a huge rangefinder. With our 58-foot beam, these 65-foot boats stuck out over each side nearly eight feet. The lumber amounted to almost three million board feet, each torpedo boat weighed fourteen tons and the rangefinder twelve tons. With such a load on deck, the ship was quite tender. I had to make sure the engineers kept the bottom tanks full at all times to maintain an even keel. After fuel was burned out of those bottom tanks they were supposed to be pumped full of sea water, "hardening up," as it's called, to keep the ship's center of gravity as low as possible.

From Puget Sound we shot straight out across the Pacific on a course for Pearl Harbor, traveling by ourselves rather than in convoy. By this time in the war some ships were allowed to do this, a lonely feeling but not so dangerous with the U.S. Navy beginning to gain more control of the Pacific.

There was a troublesome incident about halfway along the route to the Hawaiian Islands. My mate on watch sent a seaman down to my quarters one night to ask me to come up to the bridge. One of the sailors in the Navy gunnery crew had seen some flares. When I got up to the bridge, the mate told me that he hadn't seen them himself. He and I went up on the flying bridge to look in the direction the sailor said he had seen the flares. In a few moments we saw more flares, right where the young sailor said he had seen them earlier.

What to do? The Japanese were known to use ruses like this as decoys to lure merchantmen to their submarines. If it was someone really in distress I hated to leave them, but with hundreds of tons of dynamite in the holds I couldn't take a chance on coming under attack. We turned onto a new course ninety degrees from the direction we'd been sailing in, running a couple of hours in a direction away from the flares, then took a course again for Pearl Harbor.

During the next day the *Stanley Griffiths* took a decided list. I got after the engineers about hardening up those bottom tanks, but they always had some excuse. No matter what they told me, I knew their real concern was that they didn't want the trouble of cleaning the remaining sea water out of those tanks before they could be used again for fuel.

We were still listing that evening, sailing along in beautiful weather. I was standing on the bridge enjoying that weather when all of a sudden a big cloud of black smoke shot into the air from our galley. We sounded the fire alarm and ran down to the galley. The fire had been caused by a big pan of grease left on the stove by the cooks after preparing french fries for dinner. In our top-heavy condition the ship was flopping back and forth in the swells. The pan of grease had tipped over, catching fire, and the burning grease had run down into the storeroom on the deck just below the galley, setting it on fire right next to the hold loaded with dynamite. Not only did we have the fire to contend with, but it was sending a pillar of black smoke into the air that was an invitation to any Japanese submarine that might be in the area.

An emergency like that is why boat and fire drills are held. The crew put out the fire in quick order. The storeroom had a lot of smoke damage but otherwise there wasn't much harm to the ship or its cargo.

The next problem was with our refrigeration system. As the weather got warmer and warmer the temperature in the ice box started climbing. The engineers did what they could, stopping the temperature climb just close to thirty-two degrees, too warm for our frozen food to last very long. I instructed the cooks to open the door to the ice box as seldom as possible, and hoped for the best.

When we finally reached Pearl Harbor without further incident we were listed far over to one side. I had not succeeded in convincing the engineers to harden up those bottom tanks. "My God," the pilot said as he came aboard, "did you get hit by a torpedo on the way out here?" We got the ship into the dock all right, and when those torpedo boats were taken off the deck she straightened up to look like a proper ship again.

This was a trip that was plagued with problems. We had to move all our frozen food stores up to a cold storage in Pearl Harbor while the refrigeration equipment was being repaired, the cook going by taxi each day to pick up what he'd need from the cold storage. Then the engineers discovered that the boilers were in bad shape, needing to be cleaned out and re-tubed. We must have laid there a month before the work was finished and we could run to Kahului to pick up a load of canned pineapple bound for San Francisco. The weather was very bad crossing back across the Pacific, causing us to take on so much water that we had to block off the ventilators to the holds to keep the sea water from running into them. When the hatches were opened at San Francisco the holds were full of condensation brought on by running from a warm to a cold climate. There was no damage to the cans of pineapple, but the cases were a mess and I caught hell for that. No matter. It was good to get back to the States and our usual runs again.

When the pineapple mess had been cleaned up we ran over to the Howard Street Terminal in Oakland, to take on a full load of salt for the Hooker Chemical Plant in Tacoma. Discharging the salt after arrival in Tacoma took a couple of days, and then we ran to Winslow to put into the shipyard there for some general repairs. This gave me a welcome few days to spend at home with Gwen, the longest break I'd had since taking command of the *Stanley Griffiths*.

THE *STANLEY GRIFFITHS*

Once back in operation we started on a series of interesting, tough voyages to Alaska. Much of the United States aid to the Soviet Union was cargo carried on Soviet ships from Seattle to Vladivostok. Most of those ships were old coal-burners turned over to the Soviets by the British. The voyage from Seattle to Vladivostok was so long that the ships would have used up most of their cargo space to carry coal for fuel if they were not refueled somewhere along the line. To solve this problem a coaling station was set up at Akutan, in the Aleutian Islands. Our job was to haul coal from the King Street Terminal in Seattle to Akutan.

With the weather what it is in the Aleutians, and the loading procedures at Akutan, the *Stanley Griffiths* took quite a beating during the several trips we made. Sometimes we tied up at the dock in Akutan, with a Soviet vessel tied alongside us on the outside. We'd then discharge both into the other vessel and onto the dock, leaving the coal there to be picked up by another Soviet vessel. At other times we anchored out in the bay with ships moored on both sides of us to take on our load of coal.

Excerpts from the deck logs of the *Stanley Griffiths* contain quite a list of the Soviet vessels. They included the *Pakov, Jan Jorous, Rapidan, Mironych, Leningrad, Ijona, Chernyshevski, Kiev, Carl Marks, Viborg, Vtosaia Piatelctra, Askhabad, Volkhosvstrvi, Maxim Gorki, Tsielkovcki, Briansk, Sima, Evaneson, Joseph Stalin, Krasin, Lozabski, Karl Liebknecht* and *Manich*, all of them cargo ships. We also coaled a Soviet ice breaker, the *Krasnin*.

The weather was usually nasty, with heavy winds making it difficult for the Soviet vessels to tie up alongside of us. That was especially true while we were at anchor, the ship swinging on her anchor chain in the winds. Ships coming alongside usually banged into us hard. It's hard to appreciate the difficulties of operating under Aleutian Island weather conditions without having experienced them. One series of entries from the log excerpts will give some idea:

Dec. 14, 1944 2000 (hours) Russian *Yakutia* came alongside and made fast to our starboard side, coaling same from #2 hatch.

Dec. 15, 1944 0545 USSR *Yakutia* departs to bring other side to starboard.

0700 USSR *Yakutia* trying to come alongside, unable due to strong heavy wind.

1545 USSR *Yakutia* rammed ship when coming alongside and put dent in scupper that discharges from officers' toilet, also broke gangway in half on port side.

1600 Hurricane wind from East to South.

I recall one time when a Soviet vessel smashed in my lifeboats and boat davits when coming alongside. It was my duty then to submit a claim for damages to the master of the other vessel. The Russian captain wasn't very happy about that. He got out his dictionary and tried to interpret my letter. When he got to the word "damage" and realized what the letter was about, he got very excited. We called a Navy interpreter out from shore to straighten out the matter. Lieutenant Patruski, a nice young man from Boston, translated my letter into Russian for the master of the other vessel. That didn't calm him down. I found it hard to understand his excitement, considering the conditions under which he had come alongside. I didn't blame the Russian master personally for the damage, but I had a legitimate claim against his ship owners for the damages to my vessel.

The Russian's next step was to try to repair the damages with vodka. He invited me over to his cabin for a drink. We had a vodka, and then another vodka. Then they brought in crab meat and caviar. I still refused to renege on my claim. Next a couple of good-looking women were brought into the cabin, stewardesses on his ship. I didn't buy that, either. The captain was a disappointed man when I left the ship. I tried to explain that I didn't hold him at fault; it was the weather that caused his ship to crash into mine, but the law was on my side. I guess he was afraid he would catch hell when he got back home.

The captain finally found a way out the next day. Our winch driver was a little careless while swinging a coal bucket over to the other ship's bunker hatch, hitting a tripod that held a steering rod and knocking it over. A happy Russian captain immediately wrote a claims letter to me, which I signed and returned to him.

We had a friendly relationship with the crew of that ship even if the captain and I had that little set-to about the damages. After I had met the girls in his cabin they came over to our ship to ask if we had any ice cream. We did, and they were tickled when I told our steward to give them some. They were also very interested in all the magazines we had aboard, especially those that had the latest women's fashions in them.

On one of the return trips from Akutan, we went to Cold Bay to take on old ammunition, then took on more of the same at Kodiak, for delivery to Port Edwards. That is a little port near Prince Rupert, in British Columbia, where the cargo of ammunition was offloaded for rejuvenation.

Early one morning while the cargo was being discharged the mate on watch hollered "Fire!" I got up right away, to discover that we had a fire in number two hold. Ammunition piled about seven feet high still had not been offloaded from that hold. For security reasons the bridge was locked during this operation, so we ran up and unlocked it in order to blow the whistle and sound a general alarm. We got out our own hoses and the fire hoses on the dock, flooding the number two hold.

THE *STANLEY GRIFFITHS*

While all this was going on I was in the middle of an argument with the colonel in charge of operations on the dock. He ordered me to get the ship away from the dock, afraid of the damage to his installation if we blew up. I refused. We needed the fire hoses on the dock to put out the fire. The fire was put out while we stood there arguing, but that wasn't the end of it. The Army had a hearing, putting me on the carpet for a couple of days. Nothing came of it in the end, though, as they finally realized that I had no choice but to keep my ship near those big fire hoses on the dock. We never did determine the cause of the fire.

In September 1944, we were taken off the coal-hauling run to Akutan for a trip with ammunition to Adak loaded at Mukilteo. Our purser at the time was a nice young fellow named Ben Johnson. It was his first time aboard any kind of ship. Young Ben was enthusiastic, honest, and obedient, and I was looking forward to having him in my crew.

We were just about to pull up the gangway to sail from Mukilteo when the purser saw an Army lieutenant on the dock. Wanting to get a letter off to his girlfriend before we left, he handed the letter to the lieutenant and asked him if he'd mail it. The lieutenant was very obliging. Too obliging, as it turned out. The Army censored the letter, finding that Ben had told his girlfriend about our loading ammunition for Adak and giving our route to Alaska. There was a letter waiting for me at the next port, asking me to respond to some questions about Ben Johnson. I wrote a letter of high recommendation, stating I was certain that the purser would not intentionally divulge information harmful to the war effort. Ben had to go up before a Coast Guard hearing board in Seattle when we returned from this trip. We were all grateful when he was exonerated of the charges against him, as he proved to be an excellent addition to our crew.

For the next few months we made a variety of trips, sometimes running ammunition far out in the Aleutian chain in convoy from Dutch Harbor, sometimes back to coaling the Soviet ships at Akutan. There was a funny incident that winter at Attu, again involving our young purser. Some of us decided to go up to the military base to see a movie while we were laying in port there. The gunnery officer, radio operator, Ben Johnson, and I found a jeep and drove up to the base.

When we went into the theater we walked to the front to take seats there. A soldier acting as usher came to tell us that we couldn't sit there; we were in officers' uniforms and that section was reserved for enlisted men. He escorted us to the back of the theater, where there were more comfortable seats. We sat down in the row shown us by the usher, with the purser sitting in the aisle seat. The show had started when we heard a polite cough from the aisle. It was an Army general, standing alongside Ben Johnson and coughing politely to catch Ben's attention. Our gunnery officer, Lieutenant Wagner, guessed what the problem was and told Ben he was sitting in the general's seat. Ben moved, the general sat down and watched the movie with us, never saying anything about his seat being taken. Ben won a nickname for that, known thereafter as "General Johnson."

Each trip to Alaska in that winter weather was a challenge. I had a close shave at Akutan on one of those trips, when there was a full gale blowing into the harbor. It was impossible for us to stay at the dock. After we dropped anchor in the harbor it started dragging, so I dropped a second anchor. We were still dragging, getting closer to shore all the time on the port side. I decided to pull the anchors, move further out in the harbor and drop them again. When the engineers were ready I had the helmsman put the wheel hard right and called for half-ahead on the engines while heaving up on the anchors. The ship moved left instead of right, still closer to shore. I stopped right away. The wind was blowing so hard you could hardly stand up on deck, so I figured that was what was holding us from moving in the direction we wanted.

We slacked off on the anchor chains until the ship straightened out, waited for a lull in the wind and tried again. The same thing happened. By now I was really getting worried. We were very close to going on the beach to our port side. I stopped the engines and called for the second engineer to run aft to check the steering engine. He found the rudder position all the way left, due to the steering rod from the bridge having come uncoupled. I never found out how that happened. Once it was fixed we were able to move in the direction we wanted and managed to get out safely to anchor. It was one of the closest shaves I had as a ship's master.

I was to go through my closest shave not long after that, when we were at Shemya. The second-to-last port westward in the Aleutian chain, Attu being the last. Shemya was a terrible place to handle cargo. There was a dock of sorts at the port, but the harbor is open to the Bering Sea with all its awful weather. To hold at the dock we had to drop our starboard anchor, then swing around on it to back into the dock, paying out about 75 fathoms of anchor chain to act as a snubber and take strain off the lines holding us at the dock. The Army had tried to build a breakwater across the entrance to the harbor, but storms were so bad that every time they got it started another storm would come along and wash away the work that had been done.

A storm came up while we were lying at the dock, with most of the cargo unloaded. We were surging so hard against the dock that we were forced to leave before getting rid of the last part of the cargo. We started slowly ahead, taking up on the anchor chain. The anchor was stuck in the bottom of the bay, leaving us broadside to the wind and waves. From the bridge we could see the wreck of a ship on the beach. I figured they didn't need another one to join it.

We slacked off on the anchor chain, drifting back so the bow of the ship would bump against the dock in order to turn into the wind again. The gale-force winds were increasing all during this maneuver. Once straightened out, I called for a Navy steam tug that was standing by, putting his towline on our bow to keep us into the wind as we took in again on the anchor chain. We were heaving up on that anchor under a heavy strain when the chain parted suddenly at the 15-fathom mark. There was no use worrying about the loss of the anchor at that point; we had to keep going ahead to get out of that harbor and the fix we were in.

We started pumping salt water into our number four and five holds, our deep tanks and forepeak tanks, in order to get some ballast and the propeller deeper in the water. We were much too light when starting out, a condition that makes it extremely hard to hold a ship up into the wind in a storm. By the time the tug had towed us out into the Bering Sea, past the sea buoy, I figured we were in pretty good shape and let him go.

That was a mistake. The holds and tanks were still not full of enough water ballast to allow us to steer properly. The ship wouldn't hold up into the wind, falling off to the point where we were broadside to the seas, drifting towards the beach. I whistled and sent signals to the tug but he just kept going, happy to be headed back to harbor out of that storm.

At that point I decided to take a chance: to steer downwind towards the beach, trying to get up enough speed so we would have steerage way once we headed up into the wind again. I called the chief engineer up to the bridge to explain what I planned. "The reason I'm calling you here," I told him, "is to tell you not to throttle the engine. We'll need all the power we can get." Under light conditions in a heavy sea, the engineers normally throttle back on the engine when the propeller lifts out of the water, to avoid the strains caused by the propeller racing. In this case, we just couldn't afford to throttle back.

With my best helmsman on the wheel, I told him that once up into the wind he would have to be very careful to hold it there. We couldn't chance falling off again before getting well out to sea, away from that beach. The engineer and sailor both obeyed their orders. When we came around the ship held into the wind and we kept going well up into the Bering Sea before turning on a course for Adak. During that time, and while traveling to Adak, we took such a beating that the ship was heavily damaged. A lot of rivets had popped out of the forepeak, so many that it looked like a sprinkling can when we finally docked at Adak. That and some ruptured deep tanks forced us to go into drydock for repairs when we got back to Seattle.

On one of the ammunition runs to Adak that winter we loaded Navy landing barges for return to the States. Banged-up craft that had been sitting on the beach for quite awhile, they were constructed in two halves that were bolted together. We carried two of the barges dismantled into four sections, the two after sections loaded on our after deck, and the two forward sections on our forward deck. This was quite a heavy and awkward deck load. The sections were so long that they stuck out over the sides of the ship, and so high that we could barely see from the bridge over those loaded on the foredeck. Since there was no other cargo, we loaded six-hundred tons of old oil drums filled with water, putting them in the holds to lower the center of gravity.

It was late in the day when we left port. Sailing along the Bering Sea side of the Aleutians in heavy swells, I was annoyed to see that the booms were working loose and flapping around the mastheads. We'd had to top them there to make room on deck for the barges, a job I'd given to the chief mate. It was plain he'd done a poor job of it. With the ship rolling in those swells it was too dangerous to send anyone aloft on the masts to make the booms fast, so I put in at Dutch Harbor to do the job again, not very happy with my mate for his poor performance.

Light was fading when we left Dutch Harbor. I wanted to get out on the Pacific side of the Aleutian chain in daylight, so I went down by Akun Island through Akun Pass instead of through Unimak Pass. When we cleared into the Pacific all hell broke loose. A northeast gale came up, the wind blowing so hard that we couldn't stay on course. With the light load and all that surface on the landing barges for the wind to blow against, we were falling off twenty-five degrees or so from our course. To add to the problems the ship took on an eighteen-degree list. I was beginning to wish we hadn't taken so many precautions in lashing down those barges; it would have been a blessing to have them slip overboard.

That went on for a day and a half before the wind died down. Just as things were starting to get more comfortable, one of my sailors came to tell me that something was rolling around and banging inside of one of the barge sections on the after deck. I called the Navy gunnery officer and told him to go back and check what it was. He returned white as a sheet, to report that besides a lot of loose canned goods, there were a bunch of hand grenades bouncing around inside the barge. He suggested I get some of my sailors to go in there to round them up and throw them overboard. "It's your damned Navy that owns those barges," I told him, "you get your crew back there." He did, and the crew was able to toss all that stuff in the sea without an accident.

It was about two days after leaving port before I could take a sight to locate our position. I knew we had been blown well off the course we'd been ordered to follow, but I was surprised to find that we were 130 miles off our course line. With calmer winds we were able to adjust back to the original line and managed to get back to Seattle without further incident, offloading the barges at the

Navy's Pier 91.

Before our next trip we went into Winslow for some repairs. When those were completed we ran to Seattle to take on some general cargo for the Army, then to load ammunition at Mukilteo.

Because of the repairs I requested that the compass be adjusted before leaving for Alaska. A Navy lieutenant was sent aboard to carry out the adjustments.

As soon as we got to sea and the ship started rolling, the second mate called me to say something was wrong with the steering compass. Sometimes it was showing forty degrees error, other times an error of twenty degrees when compared to our standard compass. I took a look inside and found the heeling magnets in place, the magnets used to compensate for the kind of rolling the ship was doing. "I can't understand what's wrong," I told the mate.

"When the compass adjustor was working with the heeling magnets he dropped one down inside," he said.

"Didn't you make him get it out before he finished?"

No, he hadn't. I was furious. The problem was that the loose heeling magnet was rolling around at the base of the binnacle, changing the direction of the compass each time we rolled. We had an awful time lifting that binnacle off the steel deck. The screws had set up hard since the ship had been built in 1918. We finally succeeded, though, and adjusted the compass as best we could with the use of our standard compass on top of the bridge.

We arrived at Adak without further incident, where we put off a few items of cargo and took on water before proceeding to Attu, where most of the cargo was destined. We had a rough time at Attu, with so much surging at the dock that we had to go out to anchor for a time. The weather was unusually nice, though, when we left there to return to Adak. The Bering Sea was quite smooth at daybreak one morning when the mate called me up to the bridge. There was something mysterious on the horizon.

I took a look at it through the binoculars. It was a great yellow blob, impossible to describe further at that distance. We were a little nervous about it, since this was wartime and anything unusual like that was cause for concern. But since it was ahead on our course we just kept steaming along while trying to make out what it was.

The blob turned out to be one of the balloons with incendiary devices that the Japanese were sending to the western United States. They were releasing them in great numbers, to be carried by the winds to the States. There was no way they could be sent against specific targets, so they were more of a nuisance than any great danger, causing occasional forest and range fires.

The balloon we found had partly deflated, causing it to come down too soon. It was thirty or forty feet in diameter, with a big aluminum canister that could be seen through the clear water, suspended underneath. I maneuvered the ship to within about fifty feet of the balloon so we could take a good look at it, then called for the Navy gunnery officer. We talked about the possibility of bringing it aboard so the military authorities could examine and learn something from it. In the end we decided that would be too dangerous. Since it posed some hazard for navigation we decided to destroy it. I moved the ship a greater distance away, the Navy crew got their anti-aircraft gun ready, and fired. The first shell to hit the balloon was a tracer. When that tracer hit there was a huge flash of fire caused by the hydrogen filling the gas bag. The balloon disappeared.

As soon as we arrived at Adak I reported to Army Intelligence on the balloon incident. The officer who came to interview the crew agreed that we had done the right thing. He was awfully sorry, though, that we hadn't been able to bring it along so they could examine it.

We took on some cargo of military items, jeeps and the like, and proceeded from Adak to Cold Bay. I had some problems with the crew when we arrived there. The crew worked the cargo on these runs and were paid well for it. At Cold Bay we were discharging cargo picked up at Adak and Attu, working through the night according to our contract with the Army Transportation Service. About 4:30 AM my chief mate, Mr. McDaniels at the time, called to tell me that the crew had knocked off work.

I called the union delegate from the crew up to my quarters to find out what was going on. He told me they had been working sixteen hours and were all going below to sleep. I ordered the crew back to work, pointing out that they were violating their contract. The contract stated that the sailors would work cargo continuously, each man getting a four-hour break for sleep after working eight hours, then going back to work. It was a good deal for them because they received overtime pay both for the work and during their four-hour breaks. They refused to go back to work.

When we got back to Seattle I told the mate to fire the entire deck crew. They were a bad lot, had given me trouble in other ways, and after refusing a direct order to return to work during that incident in Cold Bay I had every legal right to get rid of them. The union office in Seattle howled. They refused to send me a new crew, going to the Griffiths Company and asking that I be relieved of my command. I told the company they could fire me if they wanted, but there was no way I was going back to sea with that crew. They were bad to start with, and if they won this one I'd never be able to trust them out at sea.

This incident came to an abrupt end when union officials called the Sailors Union head office in San Francisco, talking with the Union Secretary. That was Harry Lundeberg. When Harry learned that the master of the *Stanley Griffiths* was Holger Christensen, he said: "If he fired those men they deserved to be fired. Send him a new gang." And that was the end of that.

By now it was the summer of 1945. World War II was coming to an end. With a new crew we now went on a different sort of run, chartered to the Alaska Steamship Company to haul supplies for a cannery on the Nushagak River. On the way up there we had to put into Dutch Harbor for repairs to the boiler feed pumps. The pumps were rebuilt by the Navy at their installation there, and we ran as far as the mouth of the Nushagak River where I anchored to wait for a pilot. I had never been in there before. I learned then that there was just one pilot for the river, who was with the *James Griffiths* at Dillingham, that ship having arrived a day earlier. After studying the charts for awhile I figured I could make it in there on my own.

We hove anchor and started up the river. As we got near Clark Point a cannery tender came alongside, asking if we wanted a pilot. I said I'd like one, because there was a dog-leg up ahead that was little touchy. The fellow who came aboard from the tender was a man named Olson. He said he'd be happy to act as a pilot, but I should know that he didn't have a license. "How long have you been on the river?" I asked him. "Twenty-five years," he replied.

"You're the man for me. You've got the knowledge of the river, and I've got the license." He got me through that dog-leg in fine shape. I paid him fifty dollars for that, and when we finished discharging supplies at the cannery I called for him again to get us back down through there. Olson proved to be a good find.

Leaving the Nushagak River we sailed for Chernofski, in the Aleutian Islands, going back on a short contract with the Army to haul surplus Army equipment. It was a mixed load that included 77 trucks, which we took back to Tacoma. The next voyage was another commercial venture, a cargo of cannery supplies for Southeastern Alaska. After delivering the supplies to Hidden Inlet we called at Ketchikan, to load spruce lumber and other supplies destined for canneries in the westward.

While in Ketchikan I visited Captain Jack Clark, my old friend from *Griffson* days. Captain Clark's career had come together after some rocky times. Following my time with him on the *Griffson* he was a mate on the *James Griffiths*, then served as master of the *Delight*, a sister ship to the *S.A. Perkins*. He grounded the *Delight* in fog off Seattle's Magnolia Bluff, on the south side of West Point, which didn't help his career much. But in 1929 he had joined the U.S. Steamboat Inspection Service and his career went well from that point. That service was incorporated into the Coast Guard in the late 1930's. By this time he had the rank of commander, and was head of the vessel inspection division for the Coast Guard District covering southeast Alaska.

Captain Clark encouraged me to sit for the examinations to get pilot endorsements for southeast and southwest Alaska, saying that I should be certain to list all my experiences including time spent aboard the *Hannah C.* in southeast Alaska. It was good advice, and I determined to follow it.

There was a lot of miscellaneous work and calling at many ports on the remainder of this trip. The ports included Port Ashton and Port San Juan in Prince William Sound, Seldovia, Snug Harbor, Kenai, Port Graham, Port Chatham, and finally Port Bailey near Kodiak. At each of these we unloaded bits and pieces of cargo, and took on canned salmon. In addition to the canned salmon, at Kenai we had taken on a deck load of salted salmon in 350 tierces, large wooden casks of more than 100-gallon capacity. I pointed out to the cannery superintendent that I couldn't load the tierces in our holds because of all the canned salmon we were carrying, and therefore couldn't be responsible for the condition of the tierces on delivery. He signed a release form not holding the ship responsible for condition of his shipment.

By the time we left Port Bailey we had 140 thousand cases of salmon in the holds, besides all the tierces on deck. That was a heavy load, bringing the ship down to deep draft. We were heading down through Whale Passage at extreme low tide when we struck something just after passing the Ilkognak Rock Light. Whatever it was jarred the whole ship.

I stopped immediately and sounded the alarm, having everybody stand by in case of an emergency. We took soundings, checked the tanks for leaks, and generally surveyed our condition. When we could find no leaks or damage we proceeded to the Navy station at Women's Bay. I was puzzled by the fact that there should have been thirty-three feet of water where we hit, while the ship drew twenty-three feet.

Navy divers were sent down to examine the bottom of the ship. They found no damage, only a long scrape along the starboard side. I received an extract from the report of survey with the recommendation "That the vessel, being in a sound and seaworthy condition, proceed on her voyage to Seattle."

I could hardly believe my eyes when I received a follow-up letter to that recommendation, from the Sector Salvage Officer at Kodiak, stating: "Your vessel undoubtedly struck the wreck of the *Phillis S.* , operated by Bob Von Scheele of Afognak. The *Phillis S.* was cut in two just aft of the engine room by a Navy ship in December of 1942, off the entrance to Whale Pass at the point of your striking." The salvage officer had told me when we first met in Kodiak that there was a possibility I might have hit a wreck. For reasons of security, some of the notices to mariners warning of situations like this were suspended during the war.

The reason for my astonishment was that the *Phillis S.* was the boat I had served on as engineer twelve years earlier, when we nearly sank after hitting the rocks on Wada Island in the Strait of Juan de Fuca. Now, after all that time and 2,000 miles away, I had hit the wreck of that boat with a ship under my own command. The coincidence amazed me. That the *Phillis S.* survived that grounding on the rocks of Wada Island was a miracle in itself. That I should hit the wreck of that same vessel with my own ship, and sustain so little damage, was also something of a miracle.

After running to Puget Sound and discharging our big load of salmon at Seattle and Bellingham, we moved to Winslow to be dry docked for further inspection. There was more damage than the Navy divers had seen at Kodiak. A big dent in the hull required replacement of two large plates, and the straightening of several frames.

World War II had ended when we returned from this last voyage in September 1945. When repairs to the bottom of the *Stanley Griffiths* were completed, we transferred to Seattle for removal of the guns carried during wartime. Then we shifted back to Winslow to put the vessel into good condition for commercial use. She was pretty banged up from all those runs far out to the westward in Alaska, both from the weather and from the conditions for handling cargo, much of it loaded and offloaded from barges in ports where large surges made it impossible to keep the barges from bumping into us. In peacetime we might have waited for better weather in those ports before handling cargo, but in wartime, with vital military cargoes needed quickly, we took less care to avoid the kind of indents, bumps and scrapes collected during the years in that service.

It was to be almost a year before I would go to sea again. With both the *Stanley* and *James Griffiths* in the yard for repairs, the work took a long time. Those vessels were practically rebuilt, so they were almost in new condition when the work was finished. I enjoyed that time at home, the first chance to spend so many months with Gwen. It was also a profitable time for us. The captains and chief engineers remained on assignment to both ships. Besides our monthly pay we also received twenty-five dollars a day in subsistence money. Every weekday we traveled from Seattle to Winslow by ferry, to be on the ships during the day, then went back home in the evenings. All of this also gave me time to study at Captain Kildahl's Navigation School and sit for my Alaska pilot endorsements, following up on Captain Clark's suggestion.

In early January, 1946 I was assigned as master of the *Myron T. Herrick,* a ship belonging to the American President Lines. James Griffiths and Sons was acting as agent for the vessel, which was to be transferred from Seattle to San Francisco. In the end we did not go south with this ship, as the crew was trying to negotiate more pay. After a couple of weeks of arguing with the crew about their pay, the company gave up in disgust and we put her into the boneyard at Eagle Harbor. I stood by her under the same terms as with the *Stanley Griffiths* until April, when she was put into the lay-up fleet at Olympia, the first such vessel registered into that mothballed fleet of ships.

After delivering that ship to Olympia I was put back as captain of the *Stanley Griffiths*. The plans were to inaugurate a new South American run, using both the *Stanley* and *James Griffiths*. I was looking forward to command of the *James Griffiths*. The port captain for the Griffiths Company wanted command of the *Stanley Griffiths*, which was fine by me. But it was not to be. The deal being worked on by the Griffiths Company fell through, and both ships were put in the boneyard.

Seafaring men always have lots to talk about. My old friend Julius Johansson was working out his last days before retirement as captain of the *James Griffiths* at this time. On the day we were to shift both ships from the shipyard to the boneyard, he and I met at the ferry terminal in Seattle, waiting for the ferry to Winslow. We started talking about our days together, when he was the master and I the mate, getting so engrossed in our conversation that we forgot all about the ferry and missed it. I knew they couldn't move the ships without us, so I wasn't too excited, but Julius was fit to be tied. He tried to get hold of a Foss tug to take us across the Sound, with no luck. They just had to wait until we arrived much later in the morning.

Although I stood by the *Stanley Griffiths* for a while longer, that day was the last that I stood on her bridge while she was under way, the last of more than a thousand days I had served that old ship while she was in operation.

Sea Travels

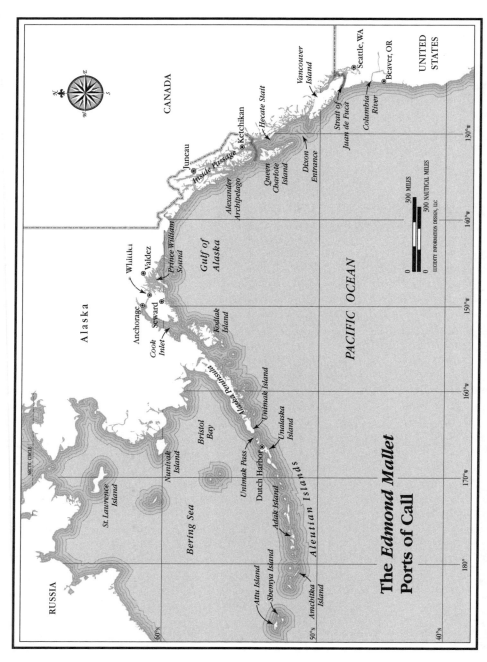

THE *EDMOND MALLET* PORTS OF CALL

CHAPTER SIXTEEN
The *Edmond Mallet*

THE *EDMOND MALLETT*

THE CONCLUSION of World War II brought an end to the heyday of American shipping. James Griffiths and Sons was no exception.

A company cannot afford to leave ships in the boneyard too long, especially ships in the top condition of the *James* and *Stanley Griffiths*. The *Stanley Griffiths* was the first to go, sold to a Swedish company. Her maiden trip for that company was to take on a load of grain at Vallejo. Fully loaded as she headed down the Sacramento River, she caught fire in the engine room. The fire boats coming to her assistance poured so much water into the engine room that the ship sank, right in the middle of the river. She was eventually raised, taken to a dry dock for repairs, and continued to sail under the Swedish flag.

In the meantime Captain Julius Johansson had retired from his last job as master of the *James Griffiths*, still in the boneyard. I was transferred to that position, with nothing to do except go over to Winslow every day to stand by her. In early October 1946 she was sold to the Chinese Government. About the same time I was called by the company's marine superintendent, Joe Sweeten, who gave me orders for a new job. This was to relieve Captain Fred Fossa as master of the *Edmond Mallet*.

After selling the *James Griffiths* the company had just two ships remaining, both under the Canadian flag. They were acting as agents for the *Edmond Mallet*, which belonged to the United States Maritime Commission and was under charter to the Army Transportation Service. She was a Liberty-class ship built in South Portland, Maine. Sometime after the original construction this ship had been reinforced with doubling plates along the entire length of both sides, riveted to the original welded plates. Seams had been cut in both wings of the deck to relieve stresses brought on by the original welding, and doubling plates were riveted over the seams. This reinforcement proved to be a lifesaver for me and my crew as we returned to the rough kind of Alaska duty experienced during my time on the *Stanley Griffiths*.

She was lying at Pier 36 in Seattle when I reported aboard, loading for Alaska with departure set for October 18th. No sooner were we loaded and ready to sail than a dispute broke out over carrying an extra pilot. We were destined for Whittier with twenty-three railway tank cars included in our deck load, where they would be put ashore to join the Alaskan railroad. The operational plan was to run in open ocean from Cape Flattery direct to Whittier. Since I had pilot endorsements for both southeast and southwest Alaska, there was no need for another pilot. But the Masters, Mates, and Pilots Union insisted that we carry one.

The dispute between the company and the union dragged on. When it appeared it would last for a while, I was ordered to take the ship to Winslow to await the outcome, mooring alongside the shipyard dock. There I was again, riding the Winslow-Seattle ferry every day to be home with Gwen during the evenings, awaiting the outcome of the dispute.

One morning when I arrived at the ship the purser came up to tell me that a fireman on the ship had died during the night. Mr. Schissler, who doubled as pharmacist's mate and purser, had called the doctor in Winslow. Dr. Tom Boren, later my own physician, came down to pronounce the man dead and issue a death certificate. I made arrangements to get the body ashore and sent into Seattle. Typical of the wide range of responsibilities borne by a ship's master, I then had to write a full report to the Coast Guard containing all details of the fireman's death.

The dispute with the union did not end until early November, when a compromise was reached. It was decided that the *Edmond Mallet* would carry mail sacks bound for Ketchikan. With Ketchikan as our first destination we would have to enter the Inside Passage. This was supposed to justify our carrying a pilot. Ironically, the pilot was with us for the remainder of the long voyage, collecting more pay than I was getting as captain. The man who came aboard in this position was an old acquaintance, Captain Hans Larson, who had such a reputation for telling tall stories that he was known as "Bullshit Hans."

Finally leaving the dock on November 6th, the first thing we did was to adjust the compass. With destinations in the Aleutian Islands, I wanted to make sure that compass was as accurate as possible. When that was accomplished we went out through the Strait of Juan de Fuca and past Cape Flattery to sail in open ocean as far as Hecate Strait, ducking into the Inside Passage there to reach Ketchikan. "You're going to have to do some work now," I told the pilot, "to earn your money." He took over the piloting there until we reached Ketchikan, three days after departure from Seattle. There was only a short stop in Ketchikan, but it gave me a chance to spend an hour over coffee with Captain Jack Clark, still assigned there with the Coast Guard. Leaving Ketchikan through *Tongass Narrows* we went up into Clarence Strait past Zarembo Island, turning around the north end of Prince of Wales Island to follow Sumner Strait out to open ocean again. On arrival at Whittier we had to wait more than a week before being able to move into the dock to discharge our cargo. The transport ships, moving military personnel in all the adjustments at the end of the war, had priority over cargo ships.

It was early December before offloading the cargo was finished. We were almost empty on the return trip to Seattle except for forty tons of general cargo that included three corpses, military men who had been killed in a railway accident in Whittier, escorted by three soldiers. I had the bodies placed in the 'tweendecks of the number four hold, due to a lesson learned when we carried some bodies south from Alaska on the *Stanley Griffiths*. On that trip I had the caskets placed in the 'tweendecks of number one hold, the hold most forward on the ship. We had a very superstitious Irishman in the deck crew at the time. He absolutely refused to stand lookout on the trip because he didn't dare pass those corpses on his way forward to the lookout station.

In Seattle we took on another load of cargo for Alaska, then sailed for Beaver, Oregon, to top off the load with ammunition bound for the Aleutian Islands. At the Columbia River Light Ship, I picked up a pilot to take us up the river. We had a little accident coming into the dock at Beaver. The pilot was cautious, but the ship took a bad swing into the dock that resulted in some smashed fender piles. Accidents like that will happen due to tides, wind and other factors affecting movement of the ship, and aren't always the fault of the pilot.

When we arrived at Beaver on December 23rd we were delayed by some crew problems that had begun in Seattle. The chief steward had quit the ship there, walking off even though he was "on articles" — the crewman's agreement to serve for a fixed period of time. The chief cook had then applied for the steward's job. Blackie Singer was a good man and had papers showing experience as a steward, so I put him in that position. Next, one of the firemen came to register a complaint about the second cook. "Captain," he said, "we won't sail unless you let the second cook go." When I asked him what the trouble was, he said that the second cook had been sleeping with one of the wipers in the engine room gang. "We don't want any fruit cooking for us," he told me. I didn't have any choice but to get rid of the second cook, leaving the ship without a proper galley crew.

Since we would have to wait in Beaver for two new cooks, I hightailed it for Seattle overland, checking in with the company and then spending the Christmas holiday with Gwen and the family. I went back to the ship right after Christmas. Soon after I got there the two new cooks appeared. Looking at them, I thought we were in for a real starvation voyage. The first cook was a little skinny guy and the second was a Filipino just as skinny. They looked like they'd been on the beach for a long time, but they turned out to be the best cooks I'd ever sailed with. The chief cook put out some wonderful meals, and the second was an excellent baker. The crew never went without delicious cinnamon rolls with their coffee. Together with Blackie Singer, who was a great chief steward, they really helped make a happy ship.

Old "B.S." Hans Larson joined the ship again as our Alaska pilot before we left Beaver. The situation with the pilots' union still had not been settled, so he once again drew a large salary with nothing to do on a long trip north. Of course we had to take on another pilot for the return trip down the Columbia, since neither "B.S." nor I had pilot endorsements for that area. Dropping the river pilot at the light ship, I shaped a course direct for Prince William Sound. We anticipated a long trip and had taken stores aboard to last for seventy-five days. As it turned out, the trip far exceeded our expectations.

We found extremely cold weather on arrival in Prince William Sound in early January, 1947. The thermometer hovered right around zero all the time we were laying in there. Steam had to be kept up on deck at all times to prevent condensation from freezing up in the steam lines and winches, which were kept turning over slowly night and day. That was a good job for the oilers. It was necessary to oil the winches regularly, and any time they came up from the engine room to do that they got overtime pay.

Since there was no berth for us in Whittier we were instructed to anchor in a nice little cove at Naked Island, a little too small for comfort. With a ship 460 feet long, and the length of anchor chain, the radius of her swing on the chain made it necessary to keep an eye out all the time to keep from putting her stern on the beach.

We lay there for many days, waiting for a berth at Whittier. There wasn't much to do in the way of ship maintenance because the ice and snow didn't allow the usual deck work. In order to get some exercise we went ashore in the ship's boats a couple of times to hike through the woods on the island. The first time ashore we hiked across the island, completely uninhabited and wild. Lo and behold, in the middle of the island we found "Kilroy was here" painted on a big rock. That guy sure got around a lot during World War II.

We were at anchor more than two weeks before word came that a dock was clear for us at Whittier. Part of our cargo was discharged there, then we left on January 23rd for Adak. It was early in the morning when we departed, the weather still extremely cold. Heading outside of Kodiak we ran into a full gale sweeping down from Cook Inlet. On the night of the 24th, off Portlock Bank, things were really tough. Extremely heavy gales and seas, and the cold weather, combined to ice down the ship.

She was a solid sheet of ice, with all the bulkheads, deck houses and machinery covered with it. We were making heavy weather, the ship rolling and laboring.

The weather moderated a little as morning broke, and by then we had gotten into the lee of Kodiak Island. Things had calmed down enough for me to send the mate down into the holds to check any damage from the storm. In number two hold we had a lot of heavy machinery loaded, including bulldozers and trucks. The bulldozers were in heavy crates when we loaded them, shored up inside with six-by-six timbers. The mate came back to report that even with that kind of shoring the bulldozers had broken the shoring inside the crates, crashed out the sides and into the trucks, smashing them all to hell. We were lucky those bulldozers didn't go right out the side of the ship.

The crew was sent down to do some more shoring and lashing so the machines wouldn't move any more. When we got into Adak I had to make a heavy-weather damage report and file a protest so insurance would take care of it. This was clearly a case of no fault to my ship and company, since the crates for the machinery all stayed where they had been lashed down in the hold, the damage coming from failure of the shoring inside the crates.

Misery loves company. On arrival in Adak I had to call a tug for assistance to dock the ship because the wind was blowing hard. When he came alongside he was going so fast that he hit us with an awful bump, putting a big dent in our shell plating. I had to call for a survey on that. The inspection showed that there were no leaks caused by the damage, just the dent in the shell plating. We had to leave it that way as there was no chance of repairing it there.

We offloaded some cargo at Adak, then sent carpenters down to number two hold to do still more shoring on that heavy equipment. Most of our cargo was bound for Attu, and except for those bulldozers and trucks it had come through that storm in good shape. Running in the Bering Sea in the middle of winter I didn't want to take a chance on any more loose cargo in the holds, so we made very sure that everything was well secured before departing on February 9th. The ship was about three-fourths loaded with cargo and in nice trim as we headed out to sea.

En route to Attu we had snow and heavy winds, increasing all the time as we got further westward along the Aleutian chain. By the night of February 11th the gale force winds were picking up to cyclonic force. Visibility had been practically nil most of the way, running along the Bering Sea side of the island chain. I had not been able to take a sight, and our radio bearings were unreliable. The stations we used were designed for aircraft navigation. The differences each time we took bearings were so great that I didn't dare rely on them. In radio contacts with Attu I was informed that the storm was expected to get still worse, so I took a course up into the Bering Sea to ride it out with plenty of sea room.

The peak of the storm hit the following morning. By then the seas were mountainous. We barely had steerage way, but even with the slowest speed possible the ship was diving into the seas with huge waves coming right over her. It's hard to estimate the size of waves in a storm like that. I do know that the bridge was fifty feet above the waterline, and that those cresting seas were higher than our position on the bridge, so they must have been fifty-five feet or more in height. One small comfort was that we weren't icing down. The ice we'd picked up in the earlier storm was from spray. In this case the ship was covered so much of the time by solid sheets of ocean water that there was no chance for it to freeze up.

In the midst of all this I got a radio message from Attu asking if we could go to the assistance of a small Navy ship. At that time we were somewhere north of Shemya Island and the Navy ship was thought to be a short distance to the east of us. I hadn't had any kind of reliable bearing for twenty-four hours and was just able to make steerage way, so I told them there was nothing we could do until the storm moderated.

It was mid-morning when the first assistant engineer came up to the bridge to tell me there was trouble with the steering engine located just below the aft deck. The forces of those terrible waves working on the rudder had put so much strain on the steering engine that three of the studs holding it down had been torn right out of the deck. I went aft with him to take a look. Those studs, made of two-inch diameter steel, sure enough had popped loose. It was apparent that the whole steering engine would go if we didn't do something about it right away.

I called the chief mate and told him to get the relieving tackle on the rudder quadrant to take the strain off the engine. He looked at me with his mouth open and told me he didn't know how to do that. I was furious. "For God's sake, do you know how to do anything?" I asked. Then I got the second mate, a fine seaman who had been a towboat man, sending him up to take care of the bridge together with the pilot. Old "B.S." Larson finally had something to do on this voyage.

I worked with the sailors to get the relieving tackle set up. It amounts to an endless double block with heavy cables run through, one set attached to each side of the rudder quadrant and then made fast to heavy padeyes fastened on the inside of the hull on each side, just opposite the quadrant. The end of each cable is then run to the cargo winches to take up the strain. When all that was hooked up it had worked to take the strain off the steering engine. I went back to the bridge.

In our next radio contact with Attu I learned that their anemometer was registering 130 knots—about 150 miles an hour, when the instrument and the building supporting it blew over and rolled down the hillside. It was a pretty impressive storm.

This was a time when I appreciated the fact that the *Edmond Mallet* had been strengthened with those doubling plates on her sides and decks.

There was still nothing we could do about helping out the Navy ship in distress. After about thirty hours of storm the sky started lightening up and the winds were moderating, so I called Attu to tell them we could now proceed to the location of the distressed vessel. I wasn't very happy about the job. Attu had informed me that the ship had anchored over a reef, which would make it difficult to put a line on her. We had been under way for about two hours and were preparing a tow line, getting things ready to help the vessel, when Attu radioed that we wouldn't be needed; a Navy tug had gone out to make the rescue.

By now we could see Shemya Island, a welcome sight. I ran around to the south side of Shemya where there was a nice bottom to anchor in the lee of the island. It was a great relief to get out of that storm about nightfall. Things had been so rough that the cooks hadn't been able to make anything more for our meals than stew, in a big pot wired to the stove. Now we were able to have a good meal and a decent night's rest. I sent a message to Attu that we were safe and would proceed into port the next morning.

We arrived at Massacre Bay about noontime on February 13th and went right in to tie up at the dock. There were still a lot of swells as an aftermath of the storm, causing us to surge at the dock. Mooring lines kept breaking. we kept putting out more and more lines, until at one point I counted twenty-seven different lines holding the ship to the dock. A survey of the ship and cargo showed surprisingly little damage to either. That extra shoring we'd put into number two hold while in Adak held through one of the worst storms I experienced in my career at sea.

It remained bitterly cold as we offloaded the cargo, with snow showers and wind that stopped the work altogether at times. Handling the cargo at Attu was a terribly slow process. With the bulk of our cargo going ashore there, with the terrible weather and a workday of only ten hours, we ended up spending more than a month at that island. Sometimes storms caused so much surging at the dock that we'd have to run out into the harbor to anchor until the weather moderated and we could again move in to the dock to continue offloading.

Despite being at such an isolated spot, my life wasn't lonely during those weeks. I made friends with Colonel Harry Forman and his aide, Captain Kim. Overall military command of Attu and Shemya was under the U.S. Navy. Colonel Forman was commander of all U.S. Army units on the islands. When we weren't out at anchor I saw him almost daily, entertaining him in my quarters or going up to parties he put on in his quarters on Attu.

During the brief times of good weather Colonel Forman and Captain Kim would come down to the ship in a jeep, inviting me to go along with them for a ride across the island. Many blacktop roads had been built there during the war and were still in good condition. I suppose most are grown over by now, but at that time we were able to drive to most any part of the island. We'd explore different areas, and sometimes Harry would take along a box of hand grenades. When he had those we'd go to Holz Bay, where the Japanese had established a base when they invaded Attu. There were still a lot of huts, dugouts and caves used by the Japanese. We used them for grenade practice, acting like a bunch of kids with fireworks.

The ship's laundry became a problem while we spent all that time in Attu. When we arrived there it was almost two months after our departure from Seattle. We had not had a chance to launder the crew's linen during that time, and it was getting cruddier and cruddier as the weeks wore on. Besides the crew's discomfort in having to use dirty sheets, the company had to pay a penalty to crew members when clean linen couldn't be provided.

The Navy had a laundry facility on Attu, operated by Army personnel. I made arrangements with Colonel Forman to put our linen ashore to be washed in that laundry, but the Navy refused to handle it when they found out it was laundry from a civilian ship. That caused a squabble between the Army and the Navy. Even though our company offered to pay for the laundry, the Navy still refused to let it be done in their facility.

Colonel Forman said he'd take care of that. He had it put aboard a power barge to be taken to Shemya, where there was a laundry under Army control.

Typical of that area in winter, the power barge ran into bad weather and had to turn around to come back to Attu with our linen still aboard. Meanwhile, I had to move the *Edmond Mallet* out to anchor again in the harbor. By then most of our cargo had been offloaded. I was advised by radio to proceed directly to Adak without returning to the dock. When I asked about our laundry, Colonel Forman sent a message: "No problem. We'll fly it to you at Adak."

We arrived at Adak a couple of days later. There was no sign of our laundry. I sent messages to Attu and Shemya but could never get a straight answer about where it was or when we could expect it. This may seem like a "great ado about nothing." It wasn't, though, considering how long our crew had gone without fresh linen and how much that can affect morale aboard. I finally sent a message to our company in Seattle, asking them to straighten out the mess. Before they could answer, I got a wire from Shemya that they would be flying our laundry to Adak. That proved to be a joy ride for some of the officers from Shemya, who rode along on the plane carrying out laundry. When a big truck carrying our linen pulled up to the ship, all the brass along on the trip came aboard to drink coffee with us. There was lots of talking about our adventures in the Aleutians.

That banner day of the clean linen was April 4, 1947. We offloaded the little remaining cargo at Adak, then took on some more cargo destined for Seward. After arriving in Resurrection Bay we went directly to the Army dock at Seward. A lieutenant in charge of security came aboard to see me when we were tied up at the dock, asking if he should place some controls on the crew to keep them from bringing bottles of liquor back to the ship. I told him no. "Our crew had a hell of a time with the weather and everything else on a long voyage out in the Aleutian Islands. Let them go ashore and get everything out of their system. When I see they've had enough I'll put a stop to it."

They sure did have a blast. Some of the sailors would have a toot during the night but sober up the next day, some never quit drinking. After three days I thought they'd had enough, so together with the chief steward and mate I went through all their quarters, confiscating all the bottles of liquor and throwing it overboard.

While laying at Seward there was a knock at the door to my quarters. I answered it to find a Navy captain in uniform. "I'm Captain Glissen from the *Tolovana*, a Navy oil tanker. Maybe you remember that I asked you to move away from my ship when we were both anchored in Attu. I came to apologize." I remembered the incident very well. He had signaled that I was too close when we anchored there, I hauled anchor to drop it further away from him, then he'd signaled that I was still too close and should move again. That time I signaled back that I'd taken bearings and was sure I had plenty of room to swing. I wouldn't move again. He had been wrong, I thought, but I told him there was no need to apologize.

Captain Glisser explained that a young lieutenant was on the bridge at the time. When he checked the bearings himself he found that I was right, so he just wanted to tell me that. I invited him in to my cabin for a drink. Afterwards we went out to dinner and had a good time together, going to a dance in town after dinner. It was a nice diversion from the steady ship's routine we'd followed for so many months.

We were all looking forward to going back to Seattle after discharging the remaining cargo at Seward. Bets among the crew were that we'd receive orders to return to Seattle light. But when the orders came they were to return to the Aleutian Islands. There was a lot of unhappiness about that. One crew member deserted, my second mate had to go to the hospital, and others were transferred. It took some time to replace crew members and take on stores for another long voyage. We finally left Seward on April 26th, bound for Adak.

It was not only the ship's crew who had been a little wild during out stay in Seward. The ship mascot, a small black spaniel, managed to sneak ashore and get herself pregnant. She was a nice dog brought on the ship before we sailed from Seattle, well cared for and liked by all the crew. Her condition became apparent as the *Edmond Mallet* steamed along in much better weather conditions than during our trek out there earlier in the year, running in ballast. We hadn't taken on fuel for all those months since leaving Seattle, so I gave the radio operator a message asking if we should bunker at Dutch Harbor or wait until arrival at Adak. We were nearing Unimak Pass and I was anxious to hear where we would refuel. When there was no reply I got hold of Sparks, learning that he hadn't yet sent the message. I told him to get right on it. I had to know if we were going in to Dutch Harbor or would be sent to Sand Bay, the refueling station at Adak.

Another hour went by without a reply. This time I went to look for Sparks myself, finding him in a prayer meeting with some other crew members. He had gotten a bad case of religion during our voyage.

"Have you got that message off yet?" I asked.

"No, Captain, this is more important. This comes first."

"This is important, all right, but so is getting fuel for the ship. Now you get in there on the key and get me that message."

I finally got him into the radio shack, pretty unhappy about having to leave his prayer meeting. He did manage to get the message out and receive a reply, ordering us to fuel at Sand Bay. But I was worried about Sparks. This incident, together with some other strange behavior, made me think he might be losing his mind.

After we had fueled at Sand Bay we moved to the dock at Adak to take on cargo. Once the cargo work was under way I went to the radio shack. "Come on, Sparks," I said, "You're going up to the doctor with me." He got wild-eyed, wondering what would happen to him, but came along with me. When he had finished taking a medical history and examining Sparks, the Army doctor informed me that he could not release Sparks for return to the ship. It turned out he had been surveyed out of the Navy for being mentally unbalanced, and would be kept at the hospital for observation.

This was just one example of the fact that a long trip under the harsh conditions we'd experienced was taking a toll on my crew. Besides having to ask for a new radio operator to be flown up from Seattle, I also had to request a replacement for one of the wipers, who was promoted to fireman due to illness of another member of the engine room gang. My casualty list would fill a book by the time we got back to Seattle. The cold weather had brought lots of cases of colds and flu, while the ice and rough seas combined to cause many accidents.

Our ship mascot helped lighten things for the crew. She was such a favorite, especially after getting pregnant, that many of them would go to see her after finishing their watches, taking food along with them. By the time she gave birth to six black pups she was fat as a butterball. We had no trouble at all in getting rid of the pups. There were many people in our various ports of call who wanted puppies, so we ended up scattering her offspring all along the Aleutian Islands.

A new radio operator and wiper arrived while we were loading Army equipment at Adak, mostly trucks and similar equipment destined for Whittier. There was a depot there where the surplus equipment used in Alaska during World War II was being gathered and inventoried for salvage. My happiness at having a full crew again disappeared during this loading process, when my winch driver suffered a double hernia. It was too close to our departure date to get a replacement, so I promoted one of the AB's to that position and sailed short one crew member.

We left Adak in early May and arrived in Attu a couple of days later to load still more surplus Army equipment. We then sailed for Amchitka, to take on more of the same from that island. That was my first time in the nice little harbor at Amchitka, reached after another two days. Fortunately, the weather was still holding good. All of those Aleutian Island harbors have a surge from the sea, even when the weather is moderate, so we seesawed against the dock as we loaded some heavy equipment. There were cement mixers, earth movers and other big items that had been used in building roads during the war.

The *Edmond Mallet* was fully loaded on departure from Amchitka at the end of May, 1947. We ran in calm but foggy weather on the Bering Sea side of the chain, going out into the Pacific through Unimak Pass. On arrival at Whittier we were able to go directly to the dock. I was treated there to my first home-cooked meal in many months by Major Fitzmorris, a very nice Army officer who invited me up to his home. His wife prepared a fine meal, and afterwards he and I shared a bottle of whiskey. I was feeling very good when he drove me back to the ship that night.

Our good luck at going directly to the Whittier dock didn't last very long. While we were still offloading surplus equipment some ships came in with higher priority cargoes, requiring us to move out to anchor in the harbor. Most of the harbor had deep water, but we found a cozy bay with a good depth for anchoring. After several days laying there, waiting for a dock to open up again, my crew was getting restless. I had the power launch put into the water so they could get out and roam around. They found a high rock near the water, where they rigged up bosun's chairs and did a fine job of painting a big sign on the rock: *"Edmond Mallet Anchorage."*

When we finally got back to the dock at Whittier I was again invited up to the Fitzsimmons' place for dinner. It had been another nice, long evening when I returned to the ship at 2:00 AM. Back aboard ship I found that all the hatches on the ship were being worked except number two, which was handled by our crew. I immediately went to find the chief mate, who told me that our sailors had knocked off because they didn't want to work anymore. Offloading was supposed to go on twenty-four hours a day, with our sailors working the number two hatch and Army soldiers handling the others. This was the same situation as I'd experienced with my crew on the *Stanley Griffiths* a couple of years earlier. They were on the same kind of contract, getting paid for four hours sleep after working eight hours, but they were also supposed to keep working until the cargo was discharged.

I was burned up. "Are the sailors supposed to tell you what to do, or are you supposed to tell the sailors?" I asked the mate. Then I went to find the colonel in charge of the dock. "If you have a gang of soldiers down in that number two hold at 8:00 o'clock in the morning, I won't let those sailors back in the hold. But I have to get a letter from you to back me up, saying that it's necessary to work the clock around." He agreed.

When the sailors found the Army working their hold the next morning they came up to raise hell with me. "You guys refused to work. I guess you don't want to work cargo," I told them. "Since you knocked off work last night you were in violation of your contract. Now you're just going to have to do ship's work, and you're not going to get any extra pay as long as we're in this port." They were a pretty unhappy bunch, growling and bitching around the ship, but I never did give in and they never received any overtime for our stay in Whittier.

Then the good news came, orders to return to Seattle. We departed Whittier on June 30th, arriving a week later to go directly to the Todd dry dock for the ship's annual inspection and overhaul. It was almost seven months since we had left Seattle for Alaska, a long trip. We were all glad to be home.

After the inspection and overhaul we moved the ship to the north side of Pier 65. There we had to have one last accident with her. The shipyard called a pilot to take her over to Pier 65. Going in through very narrow quarters, with only about ten feet to spare on each side, he drifted down on an Army transport and smashed in a liferaft and lifeboat. Even though it wasn't a very serious incident I had to make a report on it.

I stood by the ship for another week as the Griffiths Company was negotiating to buy her. That didn't work out. Next they planned to charter her, then something else, and finally she was sold to the Alaska Steamship Company. I was offered the chance of remaining aboard as master.

Gwen had been after me for years to give up the seafaring life. Since the ship had been sold out from under me, this looked like a graceful way to leave the Griffiths Company, a good time to try a new career ashore. On July 16, 1947, Captain John 0. Sellevold of the Alaska Steamship Company came aboard the *Edmond Mallet* to relieve me as master.

I was happy in a way about the decision that led to that day when I left my career at sea, unhappy that the work I'd enjoyed for so long had come to an end.

It was the hardest decision I ever made during my life.

CHAPTER SEVENTEEN
Final Voyages

DAD PASSED AWAY on the last day of 1957, the end of an uncommon life.

He did not retire from seafaring until 1946, at the age of seventy-two years. Except for those six years spent digging gold out of the beach at Nome, all of his fifty-eight years of working were spent on ships and boats. A store of sealore went with him to the grave. To me and my brothers he had passed on much knowledge and many skills, but it was only a fraction of what he must have learned in those years beginning on full-rigged sailing ships, ending with a superb knowledge of the waters of Puget Sound and southeast Alaska.

Dad's final active years, following sale of the *Hannah C.*, were spent on boats belonging to the U.S. Fish and Wildlife Service. These were boats about the size of the *Hannah C.* or bigger, used mainly for hauling crews to and from Alaska, and during the summers between many points in Alaska. His last job was as mate aboard the *Brant*, a lovely ninety-five-foot vessel that was the pride of the Fish and Wildlife Service fleet.

Before telling the highlight of the last years of Dad's career, I need to say that he was a man who was very proud of his adopted country. First-generation immigrants in the U.S. are often among the most patriotic of our citizens, and Dad would surely have been counted in this number. As told in an earlier chapter, he loved his memories of Denmark, but he was very firm in seeing to it that his children were raised as Americans with English as their first language.

Back to the *Brant*. In June of 1945, President Harry S. Truman visited Seattle. It had been just two months since he became president on the death of

Franklin D. Roosevelt in April. President Truman had inherited from Roosevelt the terrible burdens of the last year of World War II, and even during his Seattle visit the new president must have had the decision whether or not to drop atomic bombs on Japan weighing heavily on his mind. In order to give him a break from the pressures, his aides decided to take Truman salmon fishing on Puget Sound. The obvious choice of a boat was the flagship of the U.S. Fish and Wildlife Service.

It was a tremendous pride and joy for Dad to participate in taking President Truman himself out for a day on Puget Sound. The trip was written up and pictured in the newspapers, giving Dad some minor fame, but I think his real joy was the satisfaction of knowing that an immigrant son from Denmark was able to hobnob with one of the great presidents of the United States.

The master of the *Brant* was Jim Collins. Jim was a fine skipper who was a bit shy, which put Dad in the forefront with the president, other guests, and newsmen along on the fishing trip, and led to many of them believing that he was the master. Drew Pearson, a famous columnist who wrote the nationally syndicated "Washington Merry-Go-Round," included an item on this trip that mentioned Dad mistakenly as skipper and Norwegian, neither of which lessened Dad's enjoyment at seeing himself in print among such important figures as the president, Governor Mon Wallgren and Senator Warren Magnuson.

According to Pearson's account, those three played poker while the *Brant* was returning to dock in Seattle. The stakes were low, but Truman was trying hard to come out ahead. It was agreed that no matter who was ahead, the game would stop the minute the boat landed. So the President kept up a line of banter with Dad: "Slow her down, Christensen," he said, "I'm behind. I need a little extra time to catch up." Or again, it was: "Speed her up, Christensen, I'm ahead now. Let's get to shore before the governor catches up." Pearson reported that the President was a few cents ahead when they landed, and that Dad was just as pleased as Truman about that.

After his retirement Dad made his first trip to Denmark in more than half a century. I myself had the pleasure of visiting Fejø in April of 1984, when I heard many stories about the family and about Dad's visit. My favorite is that of his first encounter with a boyhood chum he hadn't seen in almost sixty years. Without showing the emotion each must have felt, they approached each other to shake hands. "Goddag, Niels," said Dad's friend. "Goddag, Hans," said Dad. And they sat down and chatted quietly just as if they had seen each other the day before.

It was a heart attack that took Dad away from us. "It's hard, growing old," he said to me when I visited the hospital shortly before the death of this man from Fejø, this man who had given me so much.

Now, in my eightieth year, I don't find it so hard to grow old as Dad did. The very active years I've spent ashore since leaving the sea prepared me better for advancing age than Dad was prepared.

After I left the *Edmond Mallet*, Gwen and I built a home on Bainbridge Island at Rockaway Beach, a home right on the waters of Puget Sound with a view into Seattle's Elliot Bay. We built a hardware store in my old home town of Winslow, which we ran successfully for many years. At the same time, I founded a corporation to construct various buildings around Winslow, the Bainbridge Island Corporation. That corporation built several of the buildings in town, including the large grocery store and an office building, and we negotiated to sell land in our mall for the post office.

I also went into partnership to buy the old Admiralty hotel in Port Townsend, the place Gwen had stayed at back in the late 1920s when she came along with us on some trips with the *Hannah C*. We sold the hardware store in 1965. I was tired of that business. Later, I went to Port Townsend to manage the hotel, buying out our partner. That was more to my liking and I enjoyed meeting the wide variety of people who stayed with us there until we sold the hotel in 1975.

But nothing in those memories can compare with those from my time at sea. I was literally raised for the sea, starting with those childhood days when Dad tethered me to the mast of *La Blanca*. I have no regrets about the years since I retired from the sea in 1947; they were good, profitable years, years in which I contributed to the community of my old home town both as a businessman and a person active in the growth of the community. If I have no regrets about those years, in all honesty I have to admit that I regret having left the sea.

Through all those years since 1947 I've kept current on my master's license, now in its ninth edition. In the early years of this period, there was always the thought in the back of my mind of someday returning to the sea. Now it's more a matter of sentiment. My current license expires in 1988. I'll probably go back to Captain Kildahl's Navigation School, still operating though the captain himself died in 1985, for a refresher course before sitting for the license.

I have had a few opportunities to be at sea during these years. The first was in July 1962, when I received an invitation from the Navy League to go along on sea trials of the aircraft carrier *Ticonderoga*, following her overhaul in the Navy shipyard at Bremerton. It was a fascinating two days and nights we spent at sea off the Washington coast, highlighted by time spent on the bridge of that huge ship. When I arrived on the bridge a three-stripe commander came towards me with a big grin on his face, holding out his hand and saying "I thought that was you when you came aboard, but had to ask one of your friends to make sure. He told me it was Captain Christensen." Commander Nielsen had been gunnery officer when I was chief mate on the *James Griffiths*. When the other officers on the bridge learned that I was an old merchant marine captain there was a lot of joking and camaraderie, most of the jokes turning on comparisons between the Navy and the merchant marines. That trip on the *Ticonderoga* was delightful for an old mariner who had been dried out by so many years spent ashore.

In 1975, I helped a friend bring a surplus Navy ship up the coast from San Diego. Boyer Halvorsen is a tow boat operator who owns several tugs and barges working mostly in Alaska, a very successful man who built up his business from a beginning with one tug boat. He had purchased a YF-class Navy ship that was put in mothballs right after World War II, a yard freighter of 130 feet in length and thirty-foot beam, powered by twin 500-horsepower diesels. Boyer bought airline tickets for me to fly to San Diego with Carl Berg, a banker with seagoing experience who would be another crew member.

We spent three or four days in San Diego before departure, getting the ship ready for the trip up the coast. The voyage turned out to be a rough one, eight days port-to-port because of very stormy weather off the coast of northern California and Oregon. After my six-hour bridge watch during the storm off Point Reyes, I turned in at midnight. When I got up at six the next morning and looked out I saw that we hadn't moved a bit, running all night into the gale with no headway made at all. That kind of weather made for a slow trip, but it was still an enjoyable experience for me to get out to sea again.

My last chance to command a vessel came in the spring and summer of 1983, when I made several trips as skipper of the *Virginia V*. A steam-powered passenger vessel, she is the last of the mosquito fleet that crisscrossed Puget Sound in the late 1800's and early part of this century, now belonging to an historical society that maintains her through profits from sightseeing cruises and excursions. My old friend, Bill Henshaw, himself retired now from a career as a ship's master and Puget Sound pilot, suggested that I join him in contributing services as captain on the various trips made by the *Virginia V*. I was happy to do it, and made several trips as her captain. It took me back many years to the time when I was a deckhand on the old *Bainbridge* as a kid. I'd probably be doing that yet if the historical society had not made different arrangements for their crews, shifting from services donated by retired mariners to a regular employment arrangement.

Navigating the yard freighter up the coast for Boyer Halvorsen, serving as captain of that historical steamer, I felt as confident during these final voyages as if I had never retired from the sea. I count myself among the lucky people who have never doubted their abilities in what they do, knowing they do it well. In looking back on my career as a mariner I realize that knowledge learned from Dad was the most important thing leading to my success, both as a seaman and as a vessel master. Those early years on *La Blanca* prepared me for my career better than any other kind of training I can imagine.

I owe my career and my own uncommon life to that man from Fejø, my father Niels Christensen, to whose memory this book is dedicated.

About the Authors

J. Holger Christensen

J. HOLGER CHRISTENSEN loved the sea, plain and simple. He learned about life at sea at an early age from the best of teachers—his father, Neils Julius Christensen, who was himself the son of a seafaring man. Neils himself had sailed from Denmark at the age of thirteen aboard one of the last tall ships.

Holger, as he preferred to be called, was the second of Neils' six children, and inherited a lifetime love of the sea. From the early days of crewing on his father's workboats, he rose from sailor to master of old cargo ships and finally World War II Liberty Ships. Holger sailed the world. His story of hard work, a love of ships, and especially the waters of the Pacific Northwest and Alaska, clearly display his passion for the sea.

Vaughn Sherman

Photo by Anna Crowley, www.annaphotography.com

VAUGHN SHERMAN was born and raised in the Seattle area, where he attended Roosevelt High School and the University of Washington. His first career was as a fisheries biologist, working in Washington State and Alaska. Following that he spent more than twenty years as an operations officer with the Central Intelligence Agency, most of the time on assignments abroad.

After retirement Vaughn was launched on a variety of community activities, many involved with the governance of nonprofit agencies and community colleges. United Way of Snohomish County included him in a Boardwalk Program that prepared volunteers as trainers for boards of United Way agencies. Through training sessions and retreat leadership he has helped many Snohomish County agencies with governance matters.

Vaughn has been deeply involved with community college education, beginning with a 1981 appointment as a trustee for Edmonds Community College. Since then he has served as President of Washington State's Trustees Association of Community and Technical Colleges (TACTC), and President of the national Association of Community College Trustees (ACCT). He is a consultant for ACCT, and has made presentations and led retreats for community college boards in all corners of the country.

He has written several articles and two monographs for ACCT and other community college organizations. He has also authored a novel based on his

CIA career, *Sasha Plotkin's Deceit,* published by Camel Press, and *Walking the Board Walk* about serving on nonprofit boards of directors, published by Patos Island Press.

A certified mediator, he is a volunteer with the Dispute Resolution Center of Snohomish, Island Counties and Skagit Counties. Vaughn lives in Edmonds with his wife, Jan Lind-Sherman, and with her enjoys a large family spread throughout the Pacific Northwest.

SEXTANT USED BY HOLGER CHRISTENSEN ON HIS SEA TRAVELS.

Other books by Vaughn Sherman

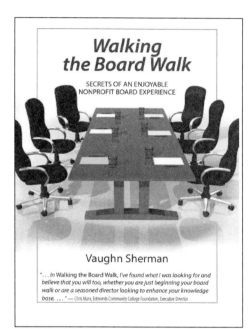

WALKING THE BOARD WALK
SECRETS OF AN ENJOYABLE NONPROFIT BOARD EXPERIENCE

An experienced trainer, negotiator, and working member of several nonprofit boards, author Vaughn Sherman will guide you and the other members of your board on how you can make your experience rewarding and enjoyable.

Find out:
- How a board works
- How the CEO and the board work together to achieve common goals
- What is expected of the each board member and the board chair
- About advocacy and fundraising

Critics' praise for *Walking the Board Walk*:

"Vaughn Sherman knows how to 'walk the board walk.' His many years of experience being on boards and conducting board trainings come through in this handy and practical guide to being an effective board for nonprofit organizations."

— Dr. Cindra Smith, author of *Trusteeship in Community Colleges: A Guide to Effective Governance (2000)*

ISBN 978-0-9847225-0-1, 96 PAGES, TRADE PAPERBACK, **$12.95**

Published by Patos Island Press

For more information, contact *www.patosislandpress.com* or *www.vaughnsherman.com*.
Distributed by *www.aftershocksmedia.com* and available at fine bookstores.

SASHA PLOTKIN'S DECEIT

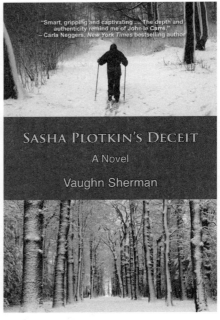

It is 1972, and the Soviet Union has succeeded in planting a mole in the top echelons of the Central Intelligence Agency. Three years earlier, CIA officer Chris Holbeck took part in a failed mission to engineer the defection of a Soviet KGB officer who may know the mole's identity. His name is Sasha Plotkin.

Chris and Sasha connected on a personal level, and Sasha indicated that he wanted to change sides. But Sasha was a no-show, and Chris would discover the full extent of Sasha Plotkin's deceit.

Years later Sasha has resurfaced and wishes to make another attempt to defect, and he wants Chris to handle it. With his wife threatening to leave and his alluring young colleague determined to seduce him, will Chris answer the call of duty? And even if Chris succeeds in getting Sasha to the United States, the question remains: Will the Soviet agent reveal the true identity of the mole?

ISBN 978-1-60381-811-7, TRADE PAPERBACK, 320 PAGES, **$16.95**

Published by Camel Press

For more information, contact *www.vaughnsherman.com*.
Distributed by *www.aftershocksmedia.com* and available at fine bookstores.